D1735637

UNIVERSITÄTS-
BIBLIOTHEK
DUISBURG
Deinventarisiert

ADVANCES IN NONLINEAR DYNAMICS AND STOCHASTIC PROCESSES

Proceedings of the meeting on nonlinear dynamics

ADVANCES IN NONLINEAR DYNAMICS AND STOCHASTIC PROCESSES

Florence 1985 (Arcetri, January 7-8)

Editors: R Livi
A Politi

Published by

World Scientific Publishing Co. Pte. Ltd.
P. O. Box 128, Farrer Road, Singapore 9128

ADVANCES IN NONLINEAR DYNAMICS AND STOCHASTIC PROCESSES

ISBN 9971-50-018-3

Printed in Singapore by Kyodo-Shing Loong Printing Industries Pte Ltd.

FOREWORD

In the recent years the community of physicists has shown a renewed and increasing interest on nonlinear dynamics and stochastic processes, both on theoretical and experimental grounds.

In Italy, as well as in other countries, one can observe a rapid growth of the number of researchers actively engaged in this field. Meetings and workshops on such topics are more frequent; this volume collects the contributions presented at the meeting on nonlinear dynamics held in Florence, January 7-8 1985.

These contributions, rather than representing pure reports of talks, are, when possible, brief reviews on the recent research carried out by various groups. The basic purpose of this meeting was to provide the participants the opportunity for comparing and discussing their results in order to favour a fruitful dialogue among different lines of research and stimulate new collaborations. The significative part of time devoted to discussions and the friendly attitude of all the participants, contributed to overcome the difficulties rising from a wide collection of topics. In fact the contributions covered a large spectrum: dynamical properties of Hamiltonian systems, soliton solutions for continuous models, onset of chaotic behaviour in dissipative systems, fluid-dynamical turbulence, stochastic differential equations, etc., which remain in the forefront of interest in theoretical and experimental physics. In particular, the analysis of bifurcation diagrams in dissipative systems, of reversible non Hamiltonian models, and of the so-called ergodic problem have been carried out both by numerical methods and new perturbative techniques. Theoretical methodologies have been applied to define new chaos indicators, to describe adiabatic invariants, and the breakdown of invariant surfaces. The connection between the KdV and the Euler equations, and the statistical mechanics for some discrete and continuum models have been studied in terms of dynamical solutions. The experimental behaviour for low (laser instabilities) and high (hydrodynamics) dimensional systems are discussed; the fractal structure of fully developed turbulence is investigated; the effect of additional stochastic forces in dissipative models is analysed theoretically and by analog simulations.

This meeting took place at the Department of Physics of the University of Florence, Arcetri. All the participants were impressed by the adverse atmospheric conditions: half a meter of snow and a terrible cold, which were not registered since half a century.

We want to thank the Department of Physics of the University of Florence and the Istituto Nazionale di Ottica in Florence for their financial and logistic support, which made this meeting possible. A special thank goes to the secretary of the meeting, Miss G. Ferasin, for her indispensable help, to Mrs. D. Scarselli and to Mrs. M.B. Petrone for their patient and careful typewriting of a large part of the manuscripts.

A particular acknowledgement goes to M. Rasetti, who warmly encouraged the production of this volume, and to Dr. K.K. Phua and all the staff of World Scientific Publishing Co. for making this issue possible.

The Editors

TABLE OF CONTENTS

ADVANCES IN NONLINEAR DYNAMICS AND STOCHASTIC PROCESSES

POINCARE'S NON - EXISTENCE THEOREM AND CLASSICAL PERTURBATION THEORY FOR NEARLY INTEGRABLE HAMILTONIAN SYSTEMS

GIANCARLO BENETTIN

Dipartimento di Fisica dell'Università di Padova
and Centro Interuniversitario di Struttura della Materia
Via Marzolo 8 – PADOVA (Italia)

LUIGI GALGANI

Dipartimento di Matematica dell'Università di Milano
Via Saldini 50 – MILANO (Italia)

ANTONIO GIORGILLI

Dipartimento di Fisica dell'Università di Milano
Via Celoria 16 – MILANO (Italia)

Abstract: *Classical perturbation theory is revisited in the light of Poincaré's theorem on the non existence of integrals of motion in nearly-integrable Hamiltonian systems.*

1. Introduction

As it is well known, at the turn of the century Poincaré [1] proved his fundamental theorem on the non-existence of integrals of motion in nearly-integrable Hamiltonian systems. Within apparently wide assumptions, this theorem states that generic perturbations modify quite significantly the basic features of integrable dynamical systems: indeed, integrals of motion disappear, and the topological character of trajectories is sensibly modified, at least in a dense subset of the phase space. On the one hand, this theorem had a quite strong impact on the development of classical mechanics as, beside its precise formulation, it was interpreted as asserting the general failure of perturbation theory for nearly-integrable Hamiltonian systems. On the other hand, because of the strong connections between absence of integrals of motion and ergodicity, the theorem, when conveniently generalized [2], was considered to support the idea that Hamiltonian dynamical systems (with at least three degrees of freedom) are in general ergodic [3], and thus suited for a statistical mechanical approach.

Today, after the basic results by Birkhoff [4], Siegel [5,6], Kolmogorov [7] and Arnold [8], and more recently Nekhoroshev [9], it is clear that Poincaré's result is not as conclusive as it was supposed. Nevertheless, Poincaré's theorem still plays a central role in classical perturbation theory, being the starting point where the different branches of classical perturbation theory come from, and consequently a basic guide to understand its modern development.

The aim of this paper is precisely to illustrate the basic methods and results of classical perturbation theory in the light of Poincaré's theorem. The paper is organized as follows: first of all, in section 2, we will introduce and prove Poincaré's theorem, although in a reduced form. Then we will shortly recall, in section 3, the Lie method for canonical transformations close to the identity, which is in our opinion the most natural and efficient one in classical perturbation theory. Then, from section 4 to section 7, we will show how, by exploiting the possibilities still left open by Poincaré's negative result, one is naturally introduced to the different branches of classical perturbation theory. In particular, section 4 is concerned with Birkhoff's construction, section 5 is devoted to Siegel's diofantine condition, section 6 deals with perturbations having a finite number of Fourier components, while section 7 introduces, within the most elementary context, the basic ideas by Arnold (the ultraviolet cut-off) and by Kolmogorov (the weak normal form).

Up to this point, only finite order, or even first order perturbation theory is considered. Instead, section 8 is devoted to infinite order perturbation theory, and reports just a sketch of Kolmogorov theorem (in the recent formulation by Poeschel [10] and by Chierchia and Gallavotti [11]) and of Nekhoroshev theorem. Finally, section 9 illustrates the generalization of Poincaré's negative result accomplished by Fermi [2], and more recently extended in ref. [12].

The reader will understand the difficulties of compressing such a wide subject in a short space, and forgive us for the frequently informal exposition.

2. Poincaré's theorem

Let us consider a nearly-integrable Hamiltonian system with n degrees of freedom. Using action-angle variables $p = (p_1, \ldots, p_n) \in \mathcal{B} \subset \mathbf{R}^n$, $q = (q_1, \ldots, q_n) \in \mathbf{T}^n$, where $\mathcal{B}$ is an open subset of $\mathbf{R}^n$ and $\mathbf{T}^n$ the n-dimensional torus, the Hamiltonian is written

$$H(p, q, \varepsilon) = h(p) + \varepsilon f(p, q, \varepsilon) \; . \tag{2.1}$$

h and f are supposed to be regular or even analytic in all of the variables. For $\varepsilon = 0$ the system is integrable, and the phase space $\mathcal{W} = \mathcal{B} \times \mathbf{T}^n$ is foliated into invariant tori, where the motion is trivial. The natural basic question is whether in general, for ε small but not zero, this property is at least partially preserved. For example, one can look for a regular canonical transformation $(p, q) = \mathcal{C}(p', q', \varepsilon)$, close to the identity, which transforms Hamiltonian (2.1) into an integrable one, i.e.

$$H'(p', q', \varepsilon) \equiv H(\mathcal{C}(p', q', \varepsilon), \varepsilon) = h'(p', \varepsilon) \; . \tag{2.2}$$

A less ambitious purpose is to give H' the so-called (nonresonant) normal form up to order r, i.e.

$$H'(p',q',\varepsilon) = h'(p',\varepsilon) + \varepsilon^{r+1} f^{(r+1)}(p',q',\varepsilon) \ , \tag{2.3}$$

for $r \geq 1$. Clearly, such a form would mean that the system behaves as an integrable one, up to times of the order of $\varepsilon^{-(r+1)}$ (more precisely, up to times smaller than $T_r = \varepsilon^{-(r+1)} \|\partial f^{(r+1)}/\partial q'\|^{-1}$, $\|.\|$ denoting a convenient norm. An even less ambitious purpose is to obtain the following weaker normal form:

$$H'(p',q',\varepsilon) = h'(p',q',\varepsilon) + \varepsilon^{r+1} f^{(r+1)}(p',q',\varepsilon) \ , \tag{2.4}$$

where h' depends only on $m < n$ angles, or more generally on $m < n$ independent linear combinations of angles. Precisely, if $\mathcal{M} \subset \mathbf{Z}^n$ is a m-dimensional module (i.e. a m-dimensional "subspace" of $\mathbf{Z}^n$), we admit

$$h'(p',q',\varepsilon) = \sum_{k \in \mathcal{M}} h'_k(p',\varepsilon) e^{ik\cdot q} \ , \tag{2.5}$$

the dot denoting the ordinary scalar product between n-tuples. The previous case corresponds to $m = 0$. From this expression, it clearly follows

$$\frac{\partial h'}{\partial q'}(p',q',\varepsilon) = \sum_{k \in \mathcal{M}} c_k(p',q',\varepsilon) k \ , \qquad c_k = i h'_k(p',\varepsilon) e^{ik\cdot q}. , \tag{2.6}$$

so that, up to time T_r, the variation of the actions is not arbitrary, being a linear combination of vectors of $\mathcal{M}$, and for any $\alpha \in \mathbf{R}$ orthogonal to $\mathcal{M}$, the quantity $\alpha \cdot p$ is an approximate integral of motion. The form (2.4), with h' given by (2.5), is called the resonant normal form, adapted to $\mathcal{M}$, up to order r.

At first sight, this program appears to be very promising: indeed, let us introduce a canonical transformation dependent on ε, for example by the standard mixed-variables method:

$$\begin{aligned} p &= p' + \varepsilon \frac{\partial S}{\partial q}(p',q,\varepsilon) \\ q' &= q + \varepsilon \frac{\partial S}{\partial p'}(p',q,\varepsilon) \\ S &= S_1(p',q) + \varepsilon S_2(p',q) + \ldots \end{aligned} \tag{2.7}$$

(any other method could be equivalently used, in particular the Lie method which will be recalled in the next section). One easily recognizes that to accomplish the program sketched above, one has to solve, at any order $r \geq 1$ in ε, one and the same equation, precisely

$$\omega(p') \cdot \frac{\partial S_r}{\partial q}(p',q) + F_r(p',q) = h'_r(p',q) \ , \tag{2.8}$$

where S_r and h'_r are unknowns, while F_r is a known term, which, for $r = 1$, is nothing but $f(p',q,0)$. Here ω denotes, as usual, the unperturbed angular frequency, i.e., $\omega = \partial h/\partial p$.

Equation (2.8) is the basic equation of classical perturbation theory; as a matter of fact, you are quite naturally led into the different branches of this theory, when you look for the possible assumptions you need, in order to be able to solve it. The starting point is, however, to establish that, in general, the above equation cannot be solved; this impossibility is indeed the heart of Poincaré's theorem, which (in a slightly reduced form) can be stated as follows:

Theorem 1 (Poincaré): *The following set of assumptions is not compatible:*

i) The unperturbed Hamiltonian $h(p)$ is strictly non-isocronous (or non-degenerate), i.e.

$$det\left(\frac{\partial^2 h}{\partial p \partial p}\right) \neq 0 \qquad \forall p \in \mathcal{B}\ ; \tag{2.9}$$

ii) The perturbation $f(p,q,0)$ has sufficiently many non-vanishing Fourier components: precisely, if

$$f(p,q,0) = \sum_{k \in \mathbf{Z}^n} f_k(p) e^{ik\cdot q}\ , \tag{2.10}$$

then for each $p \in \mathcal{B}$ and each $k \in \mathbf{Z}^n$ there exists at least a $\tilde{k} \in \mathbf{Z}^n$, $\tilde{k}$ parallel to k, such that $f_{\tilde{k}}(p) \neq 0$.

iii) There exists a canonical transformation $(p,q) = \mathcal{C}(p',q',\varepsilon)$, defined (say) by (2.7), S being regular in $\mathcal{B}' \times \mathbf{T}^n \times I$, where $\mathcal{B}'$ is an open subset of $\mathcal{B}$ and I an interval containing the origin, which gives the new Hamiltonian $H'(p',q',\varepsilon) = H(\mathcal{C}(p',q',\varepsilon),\varepsilon)$ the form

$$H'(p',q',\varepsilon) = h(p') + \varepsilon h_1'(p',q') + \varepsilon^2 f^{(2)}(p',q',\varepsilon)\ , \tag{2.11}$$

with

$$h_1'(p',q') = \sum_{k \in \mathcal{M}} h_{1k}'(p') e^{ik\cdot q}\ , \tag{2.12}$$

$\mathcal{M}$ being a module of dimension $m < n$.

Proof: by substituting (2.7) into (2.11) one finds, as already remarked, equation (2.8), with $r = 1$. One must then impose, as a necessary condition to assumption iii),

$$(ik \cdot \omega(p'))S_{1k}(p') + f_k(p') = 0 \qquad \forall k \notin \mathcal{M},\quad \forall p' \in \mathcal{B}'\ , \tag{2.13}$$

where $S_{1k}, k \in \mathbf{Z}^n$, denote the Fourier coefficients of S_1.

Now, because of i), arbitrarily close to any $p' \in \mathcal{B}'$ there exists $\tilde{p} \in \mathcal{B}'$ and $\tilde{k} \in \mathbf{Z}^n$, such that $\tilde{k} \cdot \omega(\tilde{p}) = 0$. Moreover, one can easily arrange $\tilde{k} \notin \mathcal{M}$. It follows $f_{\tilde{k}}(\tilde{p}) = 0$ as well as $f_k(\tilde{p}) = 0$ for any k parallel to $\tilde{k}$. This is in conflict with assumption ii).

Remark: Poincaré's original proof is much more complicated and subtle; it also needs the assumption that everything be analytic, while here only differentiability is used. In fact, Poincaré proves more: essentially, he is able to exclude any integral of motion independent of energy, while here only those integrals are excluded, which for $\varepsilon \to 0$ become linear combinations of actions. To our purpose, this formulation is fairly sufficient, as it stresses deeply enough the basic difficulties of classical perturbation theory.

3. Canonical transformations by the Lie method

The different possibilities to escape Poincaré's difficulties will be considered in the next sections. However, we prefer to attack the basic existence theorems of classical perturbation theory, using the Lie method to generate canonical transformations close to the identity. This method, which in our opinion is the most natural and easy for classical perturbation theory, will be here shortly recalled at a purely formal level, i.e., disregarding problems of convergence (which are the counterpart of inversion problems in the standard mixed variables canonical transformations). For more details, see ref. [13-16]; in particular, in ref. [16] all the estimates necessary to ensure convergence are explicitly performed.

The idea of the Lie method is quite simple: as it is well known, if $\chi(p,q)$ is any Hamiltonian, defined in some open domain $\mathcal{D}$, and $\mathcal{C}(p',q',t)$ denotes the solution of the corresponding Hamilton equations for initial datum (p',q'), then for any fixed t the change of variables $(p,q) = \mathcal{C}(p',q',t)$ (which will be properly defined for sufficiently small t and $(p',q') \in \mathcal{D}'$, $\mathcal{D}'$ being a suitable subset of $\mathcal{D}$), is canonical and, for t small, close to the identity. Thinking to t as a "small parameter" ε, one thus defines a family of canonical transformations $(p,q) = \mathcal{C}(p',q',\varepsilon)$ near the identity.

This canonical transformation may appear to be only implicitly defined. Instead, it is completely explicit (in particular, no inversions are needed) whenever it is used to transform functions. Indeed, let g be a regular function on $\mathcal{D}'$, and consider $(\mathcal{U}^\varepsilon g)(p',q') = g(\mathcal{C}(p',q',\varepsilon))$. One has then

$$\frac{d}{d\varepsilon}(\mathcal{U}^\varepsilon g)(p',q') = \{\chi, g\}(\mathcal{C}(p',q',\varepsilon)) \tag{3.1}$$

and consequently, for any $r > 0$ (if χ and g are sufficiently regular)

$$\mathcal{U}^\varepsilon g = \sum_{j=1}^{r} \frac{\varepsilon^r}{r!} L_\chi^r g + O(\varepsilon^{r+1}) \;, \tag{3.2}$$

with $L_\chi = \{\chi, \cdot\}$.

Let us see how this procedure works in our case, i.e. for Hamiltonian (2.1), at first order $(r=1)$. Performing one canonical transformation and retaining the terms of order one in ε, the new Hamiltonian $H'(p',q',\varepsilon) \equiv (\mathcal{U}^\varepsilon H)(p',q')$ takes the form

$$\begin{aligned} H'(p',q',\varepsilon) &= H(p',q',\varepsilon) + \varepsilon\{\chi, H\}(p',q',\varepsilon) + O(\varepsilon^2) \\ &= h(p') + \varepsilon f(p',q',0) + \varepsilon\{\chi, h\}(p',q') + O(\varepsilon^2) \;, \end{aligned} \tag{3.3}$$

and this is required to coincide up to order one with

$$h'(p',q',\varepsilon) = h(p') + \varepsilon h_1'(p',q') + O(\varepsilon^2) \;. \tag{3.4}$$

Thus one obtains $h'(p') = h(p')$ and for the unknown χ the equation

$$\{\chi, h\}(p', q') + f(p', q', 0) = h_1(p', q') \ , \tag{3.5}$$

i.e.,

$$\omega(p') \cdot \frac{\partial \chi}{\partial q}(p', q') + h_1'(p', q') = f(p', q', 0) \ , \tag{3.6}$$

which is the same as equation (2.8), with $\chi = -S$. (Notice however that the canonical transformations defined by the two methods are not identical, as they differ at higher orders in ε.)

To extend perturbation theory at any order r in ε, one could simply iterate the above procedure, setting $t = \varepsilon^2$, $t = \varepsilon^3$, and so on. A more efficient way is to consider a time-dependent Hamiltonian (generating function)

$$\begin{aligned} \chi(p, q, \varepsilon) &= \sum_{s=1}^{r} s\varepsilon^{s-1} \chi_s(p, q) \\ &= \frac{d}{d\varepsilon} \sum_{s=1}^{r} \varepsilon^s \chi_s(p, q) \end{aligned} \ , \tag{3.7}$$

defining there as before $C(p', q', \varepsilon)$ to be the solution at "time" ε of the Hamilton equations corresponding to Hamiltonian $\chi(p, q, \varepsilon)$, with initial datum (p', q'). It could be seen [13–16] that for a generic regular function $g(p, q)$ the transformed function $(\mathcal{U}^\varepsilon g)(p, q) \equiv g(C(p, q, \varepsilon))$ is explicitly produced by the following algorithm, which generalizes (3.2):

$$(\mathcal{U}^\varepsilon g)(p, q) = \sum_{s=0}^{r} \varepsilon^s g^{(s)}(p, q) + O(\varepsilon^{r+1}) \ , \tag{3.5}$$

with

$$\begin{aligned} g^{(0)}(p, q) &= g(p, q) \\ g^{(s)}(p, q) &= \sum_{j=1}^{s} \frac{j}{s} (L_{\chi_j} g^{(s-j)}(p, q) \ . \end{aligned} \tag{3.6}$$

Notice that the primes previously appearing to distinguish the new variables have been dropped: this is possible and convenient because all of the functions are defined in the same domain, while no confusion is possible, as mixed variables are never used.

By applying this algorithm to Hamiltonian (2.1), the new Hamiltonian $H' = \mathcal{U}^\varepsilon H$ is straightforwardly constructed. As it could be easily seen, if one demands

$$\begin{aligned} H'(p, q, \varepsilon) &= h(p) + \sum_{s=1}^{r} h_s'(p, q) + \varepsilon^{r+1} f^{(r+1)}(p, q, \varepsilon) \\ h_s'(p, q) &= \sum_{k \in \mathcal{M}} h_{sk}'(p) e^{ik \cdot q} \ , \end{aligned} \tag{3.7}$$

then one must solve r equations of the form

$$\omega(p) \cdot \frac{\partial \chi_s}{\partial q}(p,q) + h'_s(p,q) = F_s(p,q) \tag{3.8}$$

for the unknowns χ_s and h'_s, while the known term F_s could be explicitely produced. Clearly, F_s is obtained from $f_1, \ldots, f_s$ as well as from the already determined $h'_1, \ldots, h'_{s-1}$, $\chi_1, \ldots, \chi_{s-1}$, via a finite number of Poisson brackets.

Let us stress that everything, with this algorithm, can be explicitly constructed, and that all of the necessary estimates, leading ultimately to the estimate of the remainder $f^{(r+1)}$ in (3.7), turn out to be quite easy, as shown for example in ref. [16].

4. Birkhoff's perturbation theory

The first idea to escape Poincaré's difficulties is to modify its first assumption. Although condition (2.9) should be considered (at least in a small $\mathcal{B}$) as being generic, there are relevant cases where it is not satisfied, like the Kepler problem, or the case of weakly coupled harmonic oscillators. The latter is indeed the case considered by Birkhoff [4] (after Whittaker [17] and Cherry [18]) who assumed, in place of (2.9),

$$h(p) = \omega \cdot p \ , \tag{4.1}$$

where $\omega = (\omega_1, \ldots, \omega_n) \in \mathbf{R}^n$ is now a fixed angular frequency.

Properly speaking, Birkhoff made use of the cartesian variables $x_j = \sqrt{2p_j}\cos q_j$, $y_j = \sqrt{2p_j}\sin q_j$, $j = 1, \ldots, n$ (the ordinary positions and momenta), and considered the Hamiltonian

$$\tilde{H}(x,y) = \sum_{j=1}^{n} \frac{\omega_j}{2}(x_j^2 + y_j^2) + \sum_{r=3}^{\infty} \tilde{f}_r(x,y) \ , \tag{4.2}$$

$\tilde{f}_r(x,y)$ being a homogeneous polynomial of degree r in $(x_1, \ldots, x_n)$, $(y_1, \ldots, y_n)$. In this expression the small parameter ε does not appear explicitly, but its analog, namely the strength of the perturbation, is given by the distance to the origin, as is formally seen by the standard technique of the "blowing up". Indeed, consider the ball $\mathcal{B}_\varepsilon \in \mathbf{R}^n$ defined by $p_j = \frac{1}{2}(x_j^2 + y_j^2) < \varepsilon$, $\ j = 1, \ldots, n$, and assume, only for simplicity, that the sum in (4.2) is restricted to r even. Then, turning back to action-angle variables, and performing a trivial rescaling of the actions via a factor ε, in order to report the domain $\mathcal{B}_\varepsilon$ to $\mathcal{B}_1$, Birkhoff's Hamiltonian is converted (up to a factor ε which can be reabsorbed in a rescaling of time) into a Hamiltonian of the form

$$H(p,q,\varepsilon) = \omega \cdot p + \sum_{r=1}^{\infty} \varepsilon^r f_r(p,q) \ , \tag{4.3}$$

where $f_r(p,q)$ is homogeneous polynomial of degree $2(r+1)$ in $\cos q_1, \ldots, \cos q_n$, $\sin q_1, \ldots, \sin q_n$. So we have a standard Hamiltonian in action-angle variables of the form (2.1), with $\mathcal{B}$ given by $|p_j| < 1$, $j = 1, \ldots, n$. We thus see by the way that Birkhoff's Hamiltonian violates also hypothesis ii) of Poincaré's theorem: indeed, all of the f_r's, in particular f_1, have only a finite number of Fourier components.

Following the developments by Moser [19], Gustavson [20], and Giorgilli and Galgani [13], let us see how classical perturbation theory does work with Hamiltonian (4.3). If we use the Lie method to generate canonical transformations, we are confronted, at any order r in ε, with the equation

$$\omega \cdot \frac{\partial \chi_r}{\partial q}(p,q) + h'_r(p,q) = F_r(p,q) \; ; \tag{4.4}$$

for r=1 it is $F_1(p,q) = f(p,q,0)$.

Suppose now ω satisfies just $m \leq n-1$ independent resonance relations, i.e. that the set

$$\mathcal{M} = \{k \in \mathbf{Z}^n ;\; \omega \cdot k = 0\} \tag{4.5}$$

is a module of dimension m. Equation (4.4), for $r = 1$, is then easily seen to be solved by posing

$$\begin{aligned} h'_1(p,q) &= \sum_{k \in \mathcal{M}} f_{1k}(p) e^{ik\cdot q} \\ \chi_1(p,q) &= \sum_{k \notin \mathcal{M}} \frac{f_{1k}(p)}{ik \cdot \omega} e^{ik \cdot q} \; . \end{aligned} \tag{4.6}$$

Notice that the last sum is finite, so that there are no problems of convergence.

At higher orders in ε, the only delicate point is to recognize that F_r has always a finite number of Fourier components. This is true because, as remarked in the last section, at any order in ε, F_r is obtained via a finite number of Poisson brackets. In fact, it would not be difficult to see (for example, by meaking use of the cartesian coordinates x and y) that $F_r(p,q)$ is a homogeneous polynomial of degree $2(r+1)$ in $\cos q$ and $\sin q$. The procedure to solve equation (4.4) can then be iterated any number of times, thus accomplishing, to any order in ε, the program outlined in section 2, $\mathcal{M}$ being here imposed by the properties of the constant angular frequency ω appearing in the unperturbed Hamiltonian.

This is indeed the sketch of the proof of the following

Theorem 2 (Birkhoff): *Consider the Hamiltonian*

$$H(p,q,\varepsilon) = \omega \cdot p + \sum_{r=2}^{\infty} \varepsilon^{r-1} f_r(p,q) \; , \quad (p,q) \in \mathcal{B} \times \mathbf{Z}^n \; , \tag{4.7}$$

where $f_r(p,q)$ is a homogeneous polynomial of degree $2r$ in $\cos q$, $\sin q$, while $\omega = (\omega_1, \ldots, \omega_n)$ is a fixed angular frequency and $\mathcal{B} \in \mathbf{R}^n$ is the unit ball. Assume

$$\omega \cdot k \neq 0 \qquad \forall k \notin \mathcal{M} \; , \tag{4.8}$$

$\mathcal{M}$ being a module of dimension $m < n$.

Then, for any $r \geq 1$, and ε sufficiently small, there exists a canonical transformation generated by

$$\chi(p,q,\varepsilon) = \sum_{s=2}^{r} \varepsilon^{s-1}\hat{\chi}_s(p,q) , \tag{4.9}$$

$\hat{\chi}_s$ being a homogeneous polynomial of degree $2s$ in $\cos q$, $\sin q$, which gives the new Hamiltonian H' the form

$$H'(p,q,\varepsilon) = \omega \cdot p + \sum_{s=2}^{r} \varepsilon^{s-1}\hat{h}_s(p,q) + \varepsilon^{r+1} f^{(r+1)}(p,q,\varepsilon) \tag{4.10}$$

where $\hat{h}_s$ is also a homogeneous polynomial of degree $2s$ in $\cos q$, $\sin q$, satisfying

$$\hat{h}_s(p,q) = \sum_{k \in \mathcal{M}} \hat{h}_{sk}(p) e^{ik\cdot q} . \tag{4.10}$$

Remark: one can produce explicit expressions for both $\hat{h}_s$ and $\hat{\chi}_s$, $s = 1, \ldots, r$, and rigorous estimates for the remainder $f^{(r+1)}$.

5. An improvement to Birkhoff's program: Siegel's diofantine condition.

Let us consider again the basic equation of classical perturbation theory, namely

$$\omega \cdot \frac{\partial \chi_r}{\partial t} + h'_r = F_r . \tag{5.1}$$

As we have seen, Birkhoff's attitude against this equation is characterized by two assumptions: ω is fixed, while the known term F_r, at any r, has only finitely many Fourier components.

As pointed out by Siegel [5,6], the second condition is not necessary, as far as F_r is analytic, and ω satisfies the additional diofantine condition

$$|\omega \cdot k| \geq \gamma\, |k|^{-\eta} \qquad \forall k \in \mathbf{Z}^n , \tag{5.2}$$

where $|k| = |k_1| + \ldots + |k_n|$, while γ and η are any positive constants. As it is well known, for $\eta > n-1$ almost all frequencies satisfy this equation for some γ, although the complementary set, for any fixed γ and η, is open and dense in $\mathbf{R}^n$.

To see how these hypotheses allow solving equation (5.1), let us firstly consider the simpler equation for $\mathcal{G}(q)$

$$\omega \cdot \frac{\partial \mathcal{G}}{\partial q}(q) + \mathcal{F}(q) = 0 , \tag{5.3}$$

when the known term $\mathcal{F}(q)$ has zero average, i.e.

$$\mathcal{F}(q) = \sum_{\substack{k \in \mathbf{Z}^n \\ k \neq 0}} \mathcal{F}_k e^{ik\cdot q} , \tag{5.4}$$

and can be analitically continued for complex values of the angles, up to $|\Im q| \le \xi$, where $|\Im q| = \max_{j \le n} |\Im q_j|$, $\Im$ denoting the imaginary part of a complex number. Because of (5.4), equation (5.3) is formally solved by

$$\begin{aligned} \mathcal{G}(q) &= \sum_{\substack{k \in \mathbf{Z}^n \\ k \neq 0}} \mathcal{G}_k e^{ik \cdot q} \\ \mathcal{G}_k &= \frac{\mathcal{F}_k}{ik \cdot \omega} , \end{aligned} \tag{5.5}$$

where in virtue of (5.2) the denominators $ik \cdot \omega$ in particular never vanish.

Beside the formal level, it is quite easy to see that analyticity of $\mathcal{F}$ for $|\Im q| \le \xi$, together with the diofantine condition (5.2), imply the convergence of the Fourier series for $\mathcal{G}(q)$, in the domain $|\Im q| < \xi$. Moreover, in any domain $|\Im q| \le \xi - \delta$, where $\delta < \xi$ is any positive constant, $\mathcal{G}$ is bounded by

$$\begin{aligned} &\|\mathcal{G}\|_{\xi-\delta} \le \|\mathcal{F}\|_\xi \gamma^{-1} C_n \delta^{-2n} \\ &\max_{j \le n} \|\frac{\partial \mathcal{G}}{\partial q_j}\|_{\xi-\delta} \le \|\mathcal{F}\|_\xi \gamma^{-1} C_n \delta^{-2n-1} , \end{aligned} \tag{5.6}$$

where for any function $\mathcal{F}(q)$, analytic for $|\Im q| \le \xi$, we denote $\|\mathcal{F}\|_\xi = \sup_{|\Im q| \le \xi} |\mathcal{F}(q)|$, and C_n is a suitable constant depending only on n. The proof relies on an elementary property, characteristic of analytic functions, which is here recalled as it will be useful later too:

Technical lemma (on Fourier estimates): *If $\mathcal{F}$ is analytic in the domain $|\Im q| \le \xi$, then one has*

$$|\mathcal{F}_k| \le \|\mathcal{F}\| e^{-\xi |k|} ; \tag{5.7}$$

conversely, if $|\mathcal{G}_k| \le A e^{-\xi |k|}$, A being any positive constant, then $\mathcal{G}(q) = \sum_{k \in \mathbf{Z}^n} \mathcal{G}_k e^{ik \cdot q}$ is analytic in the domain $|\Im q| < \xi$, and for any $\delta < \xi$ it satisfies the estimates

$$\|\mathcal{G}\|_{\xi-\delta} \le A D_n e^{-\delta |k|} , \tag{5.8}$$

D_n being another n-dependent constant.

For the details, see for example ref. [8,21,22].

These considerations allow directly to solve equation (4.4) for any r: indeed, considering for example the nonresonant case $m = 0$, it is clearly sufficient to set $h'_r(p)$ equal to the average of $F_r(p,q)$ over the angles, in order to be reported to equation (5.3). The estimates (5.6) can then be used to estimate the remainder of order ε^{r+1} in the expression of the new Hamiltonian.

This idea is easily generalized to resonant frequencies, and one can prove the following more general

Lemma 3 (Siegel): *Let $\mathcal{M}'$, $\dim \mathcal{M}' = m'$, be the resonant module for ω, defined by (4.5), and assume the generalized diofantine condition*

$$|\omega \cdot k''| \geq \gamma\, |k''|^{-\eta} \qquad \forall k'' \in \mathcal{M}'' , \tag{5.9}$$

where $\mathcal{M}''$ is a suitable module such that $\mathcal{M}' \oplus \mathcal{M}'' = \mathbf{Z}^n$, and γ, η are positive constants. Let $\mathcal{F}(q)$ be analytic for $|\Im q| \leq \xi$, with no Fourier components on $\mathcal{M}'$, i.e.

$$\mathcal{F}(q) = \sum_{\substack{k \in \mathbf{Z}^n \\ k \notin \mathcal{M}'}} \mathcal{F}_k e^{ik\cdot q} . \tag{5.10}$$

Then the equation for $\mathcal{G}(q)$:

$$\omega \cdot \frac{\partial \mathcal{G}}{\partial q}(q) + \mathcal{F}(q) = 0 \tag{5.11}$$

is solved for $|\Im q| < \xi$ by

$$\mathcal{G}(q) = \sum_{\substack{k \in \mathbf{Z}^n \\ k \notin \mathcal{M}'}} \frac{\mathcal{F}_k}{ik \cdot \omega} e^{ik\cdot q} , \tag{5.12}$$

and for any $\delta < \xi$ one has the estimates

$$\begin{aligned} &\|\mathcal{G}\|_{\xi-\delta} \leq \|\mathcal{F}\|_{\xi} \gamma^{-1} C_n \delta^{-2n} \\ &\max_{j \leq n} \|\frac{\partial \mathcal{G}}{\partial q_j}\|_{\xi-\delta} \leq \|\mathcal{F}\|_{\xi} \gamma^{-1} C_n \delta^{-2n-1} , \end{aligned} \tag{5.13}$$

with a suitable n-dependent constant C_n.

Let us notice that for $m' = 0$, i.e. $\mathcal{M}' = \{0\}$, this lemma reduces to the previously sketched result.

On the basis of this lemma, equation (5.1) can be solved even for resonant frequencies, if the "projection" of ω on a suitable $\mathcal{M}''$, according to (5.9), is diofantine, and $\mathcal{F}_r$ is analytic. Indeed, it is sufficient to pose $h'_r(p,q) = \sum_{k \in \mathcal{M}'} F_{rk}(p) e^{ik\cdot q}$, in order that the lemma can be applied to find χ_r.

Birkhoff's perturbation theory can then be carried on, for small ε, up to any order r in ε, even for resonant frequencies and perturbations with infinitely many Fourier components.

6. Perturbations with finitely many Fourier components, and any $\omega(p)$.

In this section we are going to discuss the second condition entering Poincaré's theorem, precisely the requirement that the perturbation $f(p,q,0)$ in Hamiltonian (2.1) has sufficiently many Fourier components. On the contrary, no hypotheses at all will be made on $\omega(p)$, which in particular can satisfy assumption i) of Poincaré's theorem.

As pointed out already by Cherry [18], there are no obstacles to solve an equation of the form

$$\omega(p) \cdot \frac{\partial \chi}{\partial q}(p,q) + F(p,q) = h'(p,q) \tag{6.1}$$

as far as $F(p,q)$ has finitely many Fourier components, say $F_k(p) = 0$ for $|k| > K$. The only delicate point is that such an equation must be solved differently in the different regions of the phase space, according to the different properties of $\omega(p)$. This idea, which was fully exploited only ten years ago by Nekhoroshev [9], is in fact quite simple. The first region one must consider, called $\mathcal{R}_0$, is the one where one can guarantee the non-resonance condition

$$|k \cdot \omega(p)| \geq \alpha_1 \qquad \text{for } k \neq 0,\ |k| \leq K \tag{6.2}$$

α_1 being a suitable constant. For p inside this region, equation (6.1) is solved by

$$\begin{aligned} h'(p,q) &= F_0(p) \\ \chi(p,q) &= \sum_{\substack{k \in \mathbf{Z}^n \\ k \neq 0}} \frac{F_k(p)}{ik \cdot \omega(p)} e^{ik \cdot q} . \end{aligned} \tag{6.3}$$

Notice that the sum is finite, and the denominators bounded away from zero.

To solve (6.1) for p outside this region, one must consider all possible moduli $\mathcal{M} \subset \mathbf{Z}^n$, of any dimension m between 1 and n, which can be generated by m independent vectors $(k^{(1)}, \ldots, k^{(m)})$ satisfying $|k^{(j)}| \leq K$, $j = 1, \ldots, m$. Given a sequence $\alpha_1 < \ldots < \alpha_n$, to each m-dimensional module $\mathcal{M}$ one associates the region $\mathcal{R}_{\mathcal{M}}$ where one has

$$|k^{(j)} \cdot \omega(p)| < \alpha_m \tag{6.4}$$

for at least a basis $(k^{(1)}, \ldots, k^{(m)})$ of $\mathcal{M}$, with $|k^{(1)}|, \ldots, |k^{(m)}| \leq K$, while at the same time it is also (for $m < n$)

$$|k \cdot \omega(p)| > \alpha_{m+1} \qquad \forall\, k \notin \mathcal{M}, \quad |k| \leq K . \tag{6.5}$$

These regions clearly constitute a covering of the action space (notice that, depending on $\omega(p)$, some regions, including $\mathcal{R}_0$, could be empty). For details on this construction, see the paper by Nekhoroshev [9], or ref. [23]. Inside a generic region $\mathcal{R}_{\mathcal{M}}$, equation (6.1) is conveniently solved by

$$\begin{aligned} h'(p,q) &= \sum_{k \in \mathcal{M}} F_k(p) e^{ik \cdot q} \\ \chi(p,q) &= \sum_{\substack{k \in \mathbf{Z}^n \\ k \notin \mathcal{M}}} \frac{F_k(p)}{ik \cdot \omega(p)} , \end{aligned} \tag{6.6}$$

which generalize (6.3).

To have a concrete example of this quite important construction, let us consider first order perturbation theory for a Hamiltonian representing a system of n weakly coupled rotators, precisely

$$H(p,q,\varepsilon)=\sum_{j=1}^{n}\frac{p_j^2}{2}+\varepsilon\sum_{1\le i<j\le n}C_{ij}\cos(q_i-q_j)\ , \tag{6.7}$$

C_{ij} being real constants. For this Hamiltonian it is $\omega(p)=p$, while the perturbation has non-vanishing Fourier components only for $|k|\le K=2$. Models belonging to this class have been recently studied in réf. [24,25].

For such a system, the region $\mathcal{R}_0$ is defined, according to (6.2), by

$$|p_i\pm p_j|>\alpha_1 \qquad 1\le i<j\le n\ , \tag{6.8}$$

while (6.3) turns into

$$\begin{aligned} h'(p,q)&=0\\ \chi(p,q)&=\sum_{1\le i<j\le n}\frac{C_{ij}}{p_i-p_j}\sin(q_i-q_j)\ . \end{aligned} \tag{6.9}$$

Instead, if one considers, for example, the two-dimensional module generated by $k^{(1)}=(1,-1,0\ldots,0)$ and $k^{(2)}=(1,0,-1,\ldots,0)$, the corresponding region $\mathcal{R}_{\mathcal{M}}$ is defined by

$$\begin{aligned} &|p_1-p_2|,\ |p_1-p_3|\le\alpha_2\\ &|p_1+p_2|,\ |p_1+p_3|>\alpha_3\\ &|p_i\pm p_j|>\alpha_3 \qquad \text{for } (i,j)\ne(1,2),\ (1,3)\ . \end{aligned} \tag{6.10}$$

Inside this region, equation (6.1) is solved by

$$\begin{aligned} h'(p,q)&=C_{12}\cos(q_1-q_2)+C_{13}\cos(q_1-q_3)\\ \chi(p,q)&=\sum_{\substack{1\le i<j\le n\\ (i,j)\ne(1,2),\ (1,3)}}\frac{C_{ij}}{p_i-p_j}\sin(q_i-q_j)\ . \end{aligned} \tag{6.11}$$

Essentially, it has been necessary to drop two terms from the expression of χ, in order to avoid small divisors; as a consequence, two terms of the perturbation have not been killed, and still survive in h'.

Let us see how, on the basis of these considerations, perturbation theory can be carried on up to any order r as far as the perturbation $f(p,q,\varepsilon)$ appearing in Hamitonian (2.1) has finitely many Fourier components. For simplicity, we introduce the unessential restriction that f be ε independent. As usual, we give here only a sketch of the construction; for the details, see ref. [16].

The essential point is the following: the property of the perturbation of having finitely many Fourier components is preserved at any r, due to the fact that F_r is constructed via Poisson brackets of functions which, in turn, have finitely many Fourier components. More precisely, it is not difficult to recognize, on the basis of elementary algebraic considerations, that one can always guarantee $F_{rk}(p) = 0$ for $|k| > K_r \equiv rK$, if $f_k(p) = 0$ for $|k| > K$. Thus, having in mind to construct perturbation theory up to a given order $\hat{r}$, one can introduce the above decomposition of the action space, using $K_{\hat{r}}$ in place of K. Within each region one is then guaranteed that the equation

$$\omega(p) \cdot \frac{\partial \chi_r}{\partial q}(p,q) + F_r(p,q) = h'_r(p,q) \tag{6.12}$$

can be solved by (6.6), for any $r \leq \hat{r}$.

This is sufficient as far as a formal construction is considered; beyond the formal level, one can prove the following

Theorem 4 (Giorgilli and Galgani): *Consider the Hamiltonian*

$$H(p,q,\varepsilon) = h(p) + \varepsilon f(p,q) \ , \tag{6.13}$$

and assume:

i) H is analytic for $(p,q) \in \mathcal{B} \times \mathbf{T}^n$;

ii) $f_k(p) = 0$ for $|k| > K$.

Then for any $r \geq 1$, if ε is sufficiently small, one can introduce the above decomposition of the action space into regions $\mathcal{R}_{\mathcal{M}}$, with reference to $K_r = rK$, and obtain in each region, via a suitable canonical transformation $\mathcal{C}_{\mathcal{M}}(p,q,\varepsilon)$, the new Hamiltonian $H'(p,q,\varepsilon)$ in the adapted normal form

$$H'(p,q,\varepsilon) = h(p) + \sum_{s=1}^{r} \varepsilon^s h'_s(p,q) + \varepsilon^{r+1} f^{(r+1)}(p,q,\varepsilon) \ , \tag{6.14}$$

where

$$h'_s(p,q) = \sum_{k \in \mathcal{M}} h'_{sk}(p) e^{ik\cdot q} \ . \tag{6.15}$$

Remark: Notice that, by increasing r, the decomposition of $\mathcal{B}$ into resonant regions becomes finer and finer, and that h'_s, $s = 1, \ldots, r$, as well as $f^{(r+1)}$, is different for the different regions. Nevertheless, the regularity requirement, which is part of assumption iii) of Poincaré's theorem, is satisfied, at any fixed r, inside each resonant region.

7. Modifying the third of Poincaré's assumptions: Kolmogorov's and Arnold's methods

Up to now, we have seen how, by modifying either assumption i) or ii) of Poincaré's theorem, perturbation theory can be actually carried on up to any order r in ε. Here we will consider the third possibility, i.e., making weaker assumption iii). This can be done by following two basic ideas: either by looking for a weaker form of $h'(p,q,\varepsilon)$, or by accepting an essentially irregular dependence on ε. The first idea has been proposed by Kolmogorov [7], in the sketch of proof he wrote of his celebrated theorem (for a detailed proof exploiting Kolmogorov's idea, see for example ref. [22]); the second one is the so-called method of the ultraviolet cut-off, which was introduced by Arnold [9] in his own proof of Kolmogorov's theorem, and is a basic tool of classical perturbation theory.

Let us start from the latter one, and explain the basic idea in the most elementary situation. Consider equation (3.8) for $r = 1$, i.e.

$$\omega(p) \cdot \frac{\partial \chi_1}{\partial q}(p,q) + h_1'(p,q) = f(p,q) \tag{7.1}$$

(for simplicity f in (2.1) has been taken to be ε-independent). As already seen in section 5, if f is analytic for $|\Im q| \leq \xi$, then its coefficients decay exponentially with $|k|$. Then, for any $K > 0$, one can separate from f the "ultraviolet" part

$$f^{>K}(p,q) = \sum_{\substack{k \in \mathbf{Z}^n \\ |k| > K}} f_k(p) e^{ik\cdot q} , \tag{7.2}$$

which, as a consequence of (5.6), is easily seen to satisfy the estimate

$$\begin{aligned} \|f^{>K}\|_{\xi-\delta} &\leq B e^{-\frac{1}{2}\delta K} \\ B &= \left(\frac{1 + e^{-\frac{\delta}{2}}}{1 - e^{-\frac{\delta}{2}}} \right)^n \|f\|_\xi . \end{aligned} \tag{7.3}$$

In place of equation (7.1) one now solves

$$\omega(p) \cdot \frac{\partial \chi}{\partial q}(p,q) + h_1'(p,q) = f(p,q) - f^{>K}(p,q) , \tag{7.4}$$

i.e., one renounces to kill the ultraviolet part of the perturbation. But (7.4) has the same form as (7.1), with $\hat{f} = f - f^{>K}$ in place of f, so that, $\hat{f}$ having finitely many Fourier components, one can apply the method outlined in the previous section, and decompose the action space as there explained, according to the resonance properties of $\omega(p)$ with integers vectors k satisfying $|k| \leq K$. In each resonant region $\mathcal{R}_\mathcal{M}$ one then obtains

$$\begin{aligned} H'(p,q,\varepsilon) &= h(p) + \varepsilon h_1'(p,q) + \varepsilon f^{>K}(p,q) + \varepsilon^2 f^{(2)}(p,q,\varepsilon) \\ h_1'(p,q) &= \sum_{k \in \mathcal{M}} h_{1k}'(p) e^{ik\cdot q} , \end{aligned} \tag{7.5}$$

with a suitable $f^{(2)}$.

According to (7.3), if K is conveniently chosen as a function of δ and ε (a logaritmic dependence on ε^{-1} is sufficient), then the new disturbing term $\varepsilon f^{>K}$ can be made as small as one likes, say of order ε^2 as the remainder $f^{(2)}$ is. Hamiltonian (7.5) will then provide an integrable-like behavior on the appropriate time scale, exactly as Poincaré's Hamiltonian (2.11) does: however, the regularity of H' is now completely lost. Indeed, not only $f^{>K}$ depends discontinuously on ε (a fact which would be no dramatic, because it just concerns the remainder), but clearly, also the decomposition of the phase space, which strongly depends on K, will now acquire (even for fixed $r = 1$) a wild dependence on ε, as one certainly needs $K \to \infty$ for $\varepsilon \to 0$. As a result, the normalized Hamiltonian $h'(p, q, \varepsilon)$ itself will depend on ε in an essentially irregular way, against assumption iii) of Poincaré's theorem (notice that the only case where this does not happen, is that of fixed frequencies).

It could be seen that the method of the ultraviolet cut-off can be carried on up to any order r, as far as ε is sufficiently small, giving a Hamiltonian H' of the form

$$H'(p,q,\varepsilon) = h'(p,q,\varepsilon) + \varepsilon^{r+1} f_1^{(r+1)}(p,q,\varepsilon) + f_2^{(r+1)}(p,q,\varepsilon) \ , \tag{7.6}$$

where the two remainders are small for two different reasons: the former because it has ε^{r+1} in front of it; the latter because it contains ultraviolet parts, with a sufficiently large ε-dependent cut-off K.

Concerning Kolmogorov's proposal of looking for a weaker normal form, we limit ourselves to a very short exposition of the basic idea. Kolmogorov too looks, for $r = 1$, for a Hamiltonian of the form

$$H'(p,q,\varepsilon) = h'(p,q,\varepsilon) + \varepsilon^2 f^{(2)}(p,q,\varepsilon) \ ; \tag{7.7}$$

however, h' neither is integrable, nor it has (as in Poincaré's normal form) any integral of motion. Kolmogorov weak normal form is instead

$$h'(p,q,\varepsilon) = \omega(p^*)\cdot(p - p^*) + \sum_{1\le i,j\le n} a_{ij}(p,q,\varepsilon)(p_i - p_i^*)(p_j - p_j^*) \ , \tag{7.8}$$

where the parameter $p^* \in \mathcal{B}$ has the only restriction that $\omega(p^*)$ satisfies Siegel's diofantine condition (5.2). As we have seen, this condition is satisfied by a large set of frequencies, and consequently, if condition i) of Poincaré's theorem is maintained, by a large set of values of p^*.

The interest of Hamiltonian (7.8), which in virtue of (7.7) governs the dynamics of our system up to times of order ε^{-2}, is quite transparent: indeed, the torus $\{p^*\} \times \mathbf{T}^n$ is immediately checked to be invariant, and consequently, because of the arbitrariness of p^*, the original Hamiltonian $H(p,q,\varepsilon)$ admits a large set of approximately invariant tori (i.e. invariant on that time scale).

Expression (7.8) for h' is achieved by Kolmogorov via an essential use of Siegel's lemma which, as we have seen, works whenever one has fixed diofantine frequencies and analytic functions. For small ε one is guaranteed that a true (i.e. non formal) canonical transformation is defined; the disadvantage of the method is that a different canonical transformation must be introduced for each different p^*.

8. The limit $r \to \infty$: Kolmogorov-Arnold's and Nekhoroshev's results

Let us recall what we have seen up to now. On the one hand, we have considered Poincaré's non-existence theorem, which definitely makes evident the impossibility of a naive perturbation theory in classical Hamiltonian dynamics. On the other hand, we have seen that many roads nevertheless remain open, which allow one to produce different forms of classical perturbation theory. As we have sometimes commented, one can work, for sufficiently small ε, up to any finite order r. The natural question then arises whether the limit $r \to \infty$ can be performed.

A first negative result comes from a theorem by Siegel [5]: for most perturbations (in a natural topology) and any fixed ε, Biskhoff's canonical transformation diverges. Positive results come instead, as it is well known, from the theorems of Kolmogorov [7], Arnold [8] and Nekhoroshev [9].

Although proceeding on different roads, Kolmogorov's and Arnold's methods lead to the same celebrated result: if $H(p,q,\varepsilon) = h(p) + \varepsilon f(p,q)$ is analytic, $h(p)$ is strictly non isocronous, and ε is sufficiently small, then invariant tori are abundant in phase space.

We recall here a quite recent formulation of this theorem, due to Poeschel [10] and independently to Chierchia and Gallavotti [11], which is the most powerful one and at the same time the most interesting for a comparison with Poincaré's theorem.

Theorem 5 (Poeschel; Chierchia and Gallavotti): *Consider the Hamiltonian*

$$H(p,q,\varepsilon) = h(p) + \varepsilon f(p,q) \ , \tag{8.1}$$

defined for $p \in \mathcal{B} \subset \mathbf{R}^n$ and $q \in \mathbf{T}^n$, and assume :

i) H is analytic for $(p,q) \in \mathcal{B} \times \mathbf{T}^n$;

ii) $\det\left(\frac{\partial h}{\partial p \partial p}\right) \neq 0$ for $p \in \mathcal{B}$;

iii) ε is sufficiently small.

Then there exist a canonical transformation $\mathcal{C}(p,q,\varepsilon)$ and an integrable Hamiltonian $h'(p,\varepsilon)$ of class C^∞ in $\mathcal{B}_\varepsilon \times \mathbf{T}^n \times \mathcal{I}$, where $\mathcal{I}$ is an interval around the origin and $\mathcal{B}_\varepsilon$ is a subset of $\mathcal{B}$ whose border is close to the border of $\mathcal{B}$, such that the new Hamiltonian $H'(p,q,\varepsilon) = H(\mathcal{C}(p,q,\varepsilon),\varepsilon)$ satisfies the relation

$$H'(p,q,\varepsilon) \overset{\mathcal{B}_\varepsilon^*}{=} h'(p,\varepsilon) \ , \tag{8.2}$$

where $\mathcal{B}_\varepsilon^*$ is a closed set, whose measure is ε-close to the measure of $\mathcal{B}$, and $\overset{\mathcal{B}_\varepsilon^*}{=}$ denotes equality of the two members, as well as of all their derivatives, when $p \in \mathcal{B}_\varepsilon^*$.

Remark: it trivially follows from (8.2) that p is a constant of motion when it belongs to $\mathcal{B}_\varepsilon^*$; thus, there exist n integrals of motion for the original Hamiltonian (8.1), whenever the initial datum belongs to the image, through $\mathcal{C}$, of $\mathcal{B}_\varepsilon^* \times \mathbf{T}^n$. One can appreciate how thin is the separation between this result and Poincaré's non-existence theorem.

Let us come to Nekhoroshev's Theorem. This is in a sense more classical than Kolmogorov's theorem, as no "strange" sets like $\mathcal{B}_\varepsilon^*$ are involved; the proof is also more classical. While in Kolmogorov's theorem the limit $r \to \infty$ is studied at small but fixed ε, in Nekhoroshev's theorem one takes instead $r \to \infty$ and simultaneously $\varepsilon \to 0$: more precisely, one proves that r can be consistently chosen to be a convenient negative power of ε; as a consequence the remainder, which was of order ε^r, becomes exponentially small with ε, and correspondingly the time scale for which the system behaves as it were integrable becomes exponentially large.

As shown by Gallavotti [26] (see also ref. [27] and [16]), this method works straightforwardly in the case of fixed frequencies satisfying a diofantine condition, leading to the following

Theorem 6 (Nekhoroshev-Gallavotti): *Let $H(p,q,\varepsilon) = \omega \cdot p + \varepsilon f(p,q)$ be analytic for $(p,q) \in \mathcal{B} \times \mathbf{T}^n$, $\omega \in \mathbf{R}^n$ being a fixed angular frequency which, for suitable positive constants γ and η, satisfies the diofantine condition*

$$|\omega \cdot k| > \gamma |k|^{-\eta} \qquad \forall k \in \mathbf{Z}^n \, ; \quad k \neq 0 \, . \tag{8.3}$$

Then, if ε is sufficiently small, there exist constants P, T, α, β, such that any motion $\big(p(t), (q(t)\big)$, with $p(0)$ at distance at least $P\varepsilon^\alpha$ from the border of $\mathcal{B}$, satisfies the estimate

$$|p_j(t) - p_j(0)| < P\varepsilon^\alpha \qquad j = 1, \ldots, n \tag{8.4}$$

for

$$|t| \le T e^{-\left(\frac{1}{\varepsilon}\right)^\beta} \, . \tag{8.5}$$

For non isocronous systems, additional considerations of geometric nature are necessary. Indeed, at variance with Kolmogorov's case, here one wants to work in the whole of phase space. As already explained, in this case one must divide the action space into regions $\mathcal{R}_\mathcal{M}$, having well defined resonance properties, and work there separately. Clearly, this is possible for arbitrarily high times only if the orbit is somehow trapped inside these regions. Such a behavior is guaranteed by a geometric condition on $h(p)$, called by Nekhoroshev *steepness*, which is in fact nothing but a generalization of convexity. If we restrict ourselves to convex unperturbed Hamiltonians, the statement one can prove is the following

Theorem 7 (Nekhoroshev): *Consider the Hamiltonian*

$$H(p,q,\varepsilon) = h(p) + \varepsilon f(p,q) \, , \tag{8.6}$$

defined for $p \in \mathcal{B} \subset \mathbf{R}^n$ *and* $q \in \mathbf{T}^n$, *and assume :*

i) H *is analytic for* $(p,q) \in \mathcal{B} \times \mathbf{T}^n$;

ii) $\left(\frac{\partial h}{\partial p \partial p}\right)$ *is positive definite in* $\mathcal{B}$;

iii) ε *is sufficiently small.*

Then there exist constants P, T, α, β, *such that any motion* $\big(p(t),(q(t)\big)$, *with* $p(0)$ *at distance at least* $P\varepsilon^\alpha$ *from the border of* $\mathcal{B}$, *satisfies the estimate*

$$|p_j(t) - p_j(0)| < P\varepsilon^\alpha \qquad j = 1,\ldots,n \tag{8.7}$$

for

$$|t| \le Te^{-\left(\frac{1}{\varepsilon}\right)^\beta} . \tag{8.8}$$

For a more detailed statement of Nekhoroshev theorem, see the papers by Nekhoroshev [9], or ref. [23] and [27]. In the latter reference some further consequences of Nekhoroshev theorem are also drawn.

9. Extending Poincaré's non-existence result

As already remarked, the gap between Poincaré's non-existence theorem and theorem 5 is quite thin. In fact, the separation can be made even thiner, as Poincaré's result can be further extended, by a natural continuation of his arguments, as shown by Fermi [2,3] already in 1923.

The problem can be stated as follows: the presence of any regular integral of motion $I(p,q,\varepsilon)$ would imply the existence of a continuous foliation of the phase space into $(2n-1)$-dimensional invariant manifolds, of equation $I(p,q,\varepsilon) = I_0$, for I_0 in a suitable interval. The question posed by Fermi is now the following: is it possible that at least one of these manifolds — i.e., possibly even a single sheet — remains invariant for ε small but non zero? The answer, within the same assumptions of Poincaré, is negative, as far as the number of degrees of freedom exceeds two, and the manifold one is looking for is analytic. The necessity of $n > 2$ is easily understood: indeed, let ω^* be diofantine; then $\lambda\omega^*$, for $\lambda \ge 1$, is also diofantine. It follows (theorem 5) that the $(n+1)$-dimensional manifold given by $\omega(p) = \lambda\omega^*$, $\lambda \ge 1$ being a free parameter, is invariant. But for $n = 2$, it is $n+1 = 2n-1$, so that the reason why Fermi cannot exclude $(2n-1)$-dimensional invariant manifolds for $n = 2$, is just that they do exist (although, according to Poincaré's theorem, their union has empty interior). As already remarked in the introduction, the reason Fermi was so interested in the possible existence of these manifolds is the connection with the ergodic problem.

Proceeding along this direction, one can further extend this result [12], obtaining the following

Theorem 8 (Poincaré–Fermi): *Consider a nearly integrable Hamiltonian system with n degrees of freedom, with Hamiltonian*

$$H(p,q,\varepsilon) = h(p) + \varepsilon f(p,q) , \tag{9.1}$$

h and f being differentiable for $(p,q) \in \mathcal{B} \times \mathbf{T}^n$. Assume:

i) $\det\left(\frac{\partial^2 h}{\partial p \partial p}\right) \neq 0$ *for* $p \in \mathcal{B}$;

ii) the perturbation $f(p,q)$ has sufficiently many Fourier components, as in Poincaré's assumption ii);

iii) there exists a m-dimensional invariant manifold V, $m < 2n$, defined by $2n - m$ equations of the form

$$I_j(p,q,\varepsilon) = 0 \qquad j = 1,\ldots,2n-m , \tag{9.2}$$

with $I_j(p,q,\varepsilon) = I_j^{(0)}(p) + \varepsilon I_j^{(1)}(p,q,\varepsilon)$, the gradients of $I_j^{(0)}$, $j = 1,\ldots,2n-m$ and the gradient of $h(p)$ being linearly independent.

Then V must be exactly $(n+1)$-dimensional, and defined by an equation of the form $\omega(p) = \lambda\omega^*$, where λ is a free parameter (within a suitable range), and ω^* is a non-resonant angular frequency.

Remark: to exclude manifolds of dimension $m = 2n - 1$, the assumption that $\operatorname{grad} I^{(0)}$ and $\operatorname{grad} h$ be linearly independent can be replaced (as Fermi did) by analyticity of both H and I in all variables.

After this result, we can say that Poincaré's non-existence arguments, and modern existence theorems, essentially touch each other.

REFERENCES

[1] H. Poincaré: *Les Méthodes Nouvelles de la Méchanique Céleste*, Vol. 3 (Gautier–Villars, Paris, 1899).

[2] E. Fermi: *Nuovo Cimento* **26**, 105 (1923).

[3] E. Fermi: *Nuovo Cimento* **25**, 267 (1923); *Phys. Z.*, **24** (1923)

[4] G. D. Birkhoff: *Dynamical Systems*, (Am. Math. Soc., New York 1927).

[5] C. L. Siegel: *Ann. Math.*, **43**, 607 (1942).

[6] C. L. Siegel and J. K. Moser: *Lectures on Celestial Mechanics* (Springer Verlag, Hidelberg 1971).

[7] A. N. Kolmogorov: *Dokl. Akad. Nauk. SSSR* **98**, 527 (1954); English translation in G. Casati and G. Ford (Editors): *Lecture Notes in Physics* No. 93 (Springer Verlag, Berlin, 1979).

[8] V. I. Arnold: *Usp. Mat. Nauk* **18**, 13 (1963) [*Russ. Math. Surv.* **18**, 9 (1963)]; *Usp. Mat. Nauk* **18**, 91 (1963) [*Russ. Math. Surv.* **18**, 85 (1963)].

[9] N. N. Nekhoroshev: *Usp. Mat. Nauk* **32**, (1977) [*Russ. Math. Surv.* **32**, 1 (1977)]; *Trudy Sem. Petrows.* No. 5, 5 (1979).

[10] J. Poeschel: *Commun. Pure Appl. Math.* **35**, 653 (1982); *Celestial Mechanics* **28**, 133 (1982).

[11] L. Chierchia and G. Gallavotti: *Nuovo Cimento B* **67**, 277 (1982).

[12] G. Benettin, G. Ferrari, L. Galgani and A. Giorgilli: *Nuovo Cimento B* **72**, 137 (1982).

[13] A. Giorgilli and L. Galgani: *Celestial Mechanics* **17**, 267 (1978).

[14] J. Henrard: *Celestial Mechanics* **3**, 107 (1970); *Celestial Mechanics* **10**, 497 (1974).

[15] M. Rapaport: *Celestial Mechanics* **28**, 291 (1982).

[16] A. Giorgilli and L. Galgani: *Rigorous Estimates for the Series Expansions of Hamiltonian Perturbation Theory*, preprint.

[17] E. T. Whittaker: *Proc. Roy. Soc. Edinb.* **A37**, 95 (1916).

[18] T. M. Cherry: *Proc. Cambr. Phil. Soc.* **22**, 287, 325 and 510 (1924).

[19] J. Moser: *Lectures on Hamiltonian Systems*, in *Mem. Am. Math. Soc.* No 81, 1 (1968).

[20] F. Gustavson: *Astron. J.* **71**, 670 (1966).

[21] G. Gallavotti: *The Elements of Mechanics* (Springer Verlag, Berlin 1983).

[22] G. Benettin, L. Galgani, A. Giorgilli and J.-M. Strelcyn: *Nuovo Cimento B* **79**, 201 (1984).

[23] G. Benettin, L. Galgani, A. Giorgilli: *A Proof of Nekhoroshev's Theorem for the Stability Times in Nearly Integrable Hamiltonian Systems*, to appear in *Celestial Mechanics* .

[24] G. Benettin, L. Galgani, A. Giorgilli: *Classical Perturbation Theory for a System of Weakly Coupled Rotators*, and: *Numerical Investigation on a Chain of Weakly Coupled Rotators in the Light of Classical Perturbation Theory*, both to appear in *Nuovo Cimento B*.

[25] C. E. Wayne: *On the Elimination of Non-Resonant Harmonics* and *Bounds on Trajectories of a System of Weakly Coupled Rotators*, to appear in *J. Stat. Phys.*

[26] G. Gallavotti: lectures given at the 1984 *Les Houches* Summer School, to be published.

[27] G. Benettin and G. Gallavotti: *Exponential Estimates for the Stability times in Nearly Integrable Hamiltonian Systems*, to appear in *J. Stat. Phys.*

EQUIPARTITION PROBLEM IN THE THERMODYNAMIC LIMIT AND ULTRAVIOLET CATASTROPHE IN CLASSICAL FIELD THEORY

Roberto Livi*, Marco Pettini°, Stefano Ruffo*, Angelo Vulpiani"

* Dipartimento di Fisica dell'Università degli Studi di Firenze and Istituto Nazionale di Fisica Nucleare
Largo E. Fermi 2, I-50125 Firenze, Italy.

° Osservatorio Astrofisico di Arcetri and Gruppo Nazionale di Astronomia del Consiglio Nazionale delle Ricerche,
Largo E. Fermi 5, I-50125 Firenze, Italy.

" Dipartimento di Fisica, Università "La Sapienza" and Gruppo Nazionale di Struttura della Materia del Consiglio Nazionale delle Ricerche,
Piazzale A. Moro 2, I-00185 Roma, Italy.

(Dedicated to Massimo Sparpaglione, our friend)

ABSTRACT

It is shown that both the thermodynamic and the continuum limit do not necessarily imply the equipartition of energy. In the latter case equipartition is incompatible with analiticity properties of the equations of motion.
We study some applications to models of weakly nonlinear solids (Fermi-Pasta-Ulam model) and of a radiant cavity.

Since the first numerical experiment on weakly coupled oscillators performed by Fermi, Pasta and Ulam (F.P.U.) (1) it was evident that for hamiltonian systems at low energy there is not equipartition of the energy among the degrees of freedom and that ergodicity is broken. This result was in complete disagreement with the idea (usually assumed by the community of physicists) that the ergodic problem was substantially solved. This opinion was based on a wrong interpretation of a theorem of Poincaré and Fermi on the non existence of prime integrals independent from the energy in a generic hamiltonian system.

The scenario of the breakdown of ergodicity (and equipartition) has been confirmed theoretically (essentially by K.A.M. theorem) (2) and in a huge amount of numerical experiments (3). Roughly speaking in a generic hamiltonian system with N degrees of freedom there is a "critical" energy E_c so that for $E \lesssim E_c$ there is a prevalence of ordered orbits while for

$E \gtrsim E_c$ the chaotic trajectories are dominant.

A relevant problem for the foundations of classical statistical mechanics is the behaviour of E_c at large N, i.e. if $\lim_{N\to\infty} \frac{E_c}{N}$ is zero or a finite value. In the "soviet" school (4) the dominant idea is that $\frac{E_c}{N} \xrightarrow[N\to\infty]{} 0$; on the contrary in the "italian" school (3) (Scotti, Galgani, ...) the prevailing idea is that $\frac{E_c}{N} \xrightarrow[N\to\infty]{}$ const > 0 and that the persistence of ordered motions at low energy density also for $N \gg 1$ is relevant for the foundations of classical statistical mechanics. The violation of theorems based on the existence of a uniform invasion of the phase space, e.g. the equipartition theorem, is a consequence of the persistence of ordered motions.

If one wants to try to understand numerically the problem of the breakdown of equipartition in the limit $N \gg 1$ one is usually faced with two problems:

a) clarify the concept of critical energy;
b) find an "equipartition" indicator easy to compute and such that it gives good informations on the states of the system.

Indeed the concept of "chaoticity" is very vague and usually it means exponential divergence of initially closed orbits (i.e. positive maximum Lyapunov exponent), continuous power spectra, strong energy sharing among the normal modes. Moreover, as Benettin et al. (3) noticed in general cases, chaoticity (in the sense of positive maximum Lyapunov exponent) does not imply equipartition: it is possible that the phase space is divided in two or more regions, inside which the motion is chaotic; therefore ergodicity (and equipartition) is broken.

We have chosen to work on the equipartition problem for a chain of coupled particles whose hamiltonian is given by

$$H = \sum_{i=1}^{N} \left\{ \frac{1}{2} \left[p_i^2 + (q_{i+1} - q_i)^2 \right] + \lambda V(q_{i+1} - q_i) \right\} , \quad q_1 = q_{N+1} \tag{1}$$

where q_i are the displacements about the equilibrium positions of the particles; p_i are the conjugate momenta and $V(\cdot)$ is a non quadratic polynomial potential. In order to study the equipartition energy density threshold, ε_c in the limit $N \gg 1$ we start with an initial condition very far from the equipartition state:

$$q_i(0) = \sum_{n=1}^{N/2} \left\{ A_n(0) \cos\left(\frac{2\pi n}{N}(i-1) \right) + B_n(0) \sin\left(\frac{2\pi n}{N}(i-1) \right) \right\} \tag{2}$$

$$p_i(0) = 0$$

with the assumption that the only non-vanishing $A_n(0)$, $B_n(0)$ terms are those with $n \in |\bar{n},\bar{n}+(\Delta n-1)|$. The initial conditions above are chosen in such a way that at t=0 the energy of the system is uniformly distributed only among Δn normal modes with wavenumbers $\bar{n}$, $\bar{n}+1,\ldots,\bar{n}+(\Delta n-1)$.

The mode-mode coupling, due to the nonlinear terms in the equation of motion, leads to energy sharing among all the normal modes. In order to observe whether the system approaches equipartition or not we have introduced a quantity that gives a distance of the asymptotic state from the equipartition state. To define this quantity (spectral entropy) we introduce at any time the Fourier decomposition:

$$[3] \quad q_i(t) = \sum_{n=1}^{N/2} \left\{ A_n(t) \cos\left(\frac{2\pi n}{N}(i-1)\right) + B_n(t) \sin\left(\frac{2\pi n}{N}(i-1)\right) \right\}$$

so that the harmonic energy of each normal mode is

$$[4] \quad E_n(t) = \frac{1}{2}\left\{ \dot{A}_n(t)^2 + \dot{B}_n(t)^2 + \omega_n^2 \left[A_n(t)^2 + B_n(t)^2 \right] \right\}$$

where $\omega_n = 2 \sin\left(\frac{\pi n}{N}\right)$ is the frequency of n-th normal mode in the harmonic approximation. Now we define $P_n(t)$ as

$$[5] \quad P_n(t) = \frac{\langle E_n(t) \rangle_T}{\sum_{k=1}^{N/2} \langle E_k(t) \rangle_T}$$

where $\langle E_n(t) \rangle_T$ is defined as

$$[6] \quad \langle E_n(t) \rangle_T = \frac{1}{T} \int_{t-\frac{T}{2}}^{t+\frac{T}{2}} dt' \, E_n(t')$$

T is chosen much larger than the lowest harmonic period in order to eliminate short time fluctuations. Now we can introduce the spectral entropy:

$$[7] \quad S(t) = -\sum_{n=1}^{N/2} P_n(t) \ln P_n(t)$$

Clearly the "asymptotic" (numerically read large times) value S_A of S gives a measure of the degree of equipartition. Indeed S_A has a maximum value $S_{MAX} = \ln\frac{N}{2}$ when $P_n = 2/N$ (i.e. equipartition) while in the harmonic limit there is no energy sharing and $S_A = S(0) = \ln(\Delta n)$. In order to have a quantity which no longer suffers the N-dependence of the maximum value S_{MAX} we define a normalized spectral entropy.

$$[8] \qquad \eta(t) = (S_{MAX} - S(t)) / (S_{MAX} - S(0))$$

Note that η varies between 1, i.e. perfect harmonicity, and 0, i.e. complete equipartition of energy.

We have performed numerical experiments at different values of N taking $\bar{n} \propto N$ and $\Delta n \propto N$ so that the wavelenght of the lowest excited mode remains constant as well as the density $\Delta n/_N$ of the initially excited modes. These choices are introduced in order to mimic the thermodynamic limit, where the number of particles and the volume both increase with a finite constant ratio.

The results for the F.P.U. ß model, i.e. $V(\xi) = \frac{1}{4}\xi^4$ are shown in Fig.(1). It is evident that ε_c (i.e. the value of the energy density such that for $\varepsilon > \varepsilon_c$ there is a complete equipartition) is independent on N.

A quite natural problem is to understand whether the result obtained for the F.P.U. ß-model is a peculiar property of this system or a generic one. To check the independence of ε_c on N we have perturbed the F.P.U. ß-model with random quenched fluctuations of the nonlinear coupling constant, i.e. we have considered the hamiltonian:

$$[9] \qquad H = \sum_{i=1}^{N} \left\{ \frac{1}{2}\left[p_i^2 + (q_{i+1} - q_i)^2\right] + \frac{\beta_i}{4}(q_{i+1} - q_i)^4 \right\}$$

where β_i are random quenched variables with a given mean value ß. We found that the two systems with $\beta_i = \beta$ and random β_i exibit the same statistical behaviour.

Another test has been done considering the so called F.P.U. α-model:

$$V(\xi) = \frac{1}{3}\xi^3$$

The results are shown in Fig. (2). Similar results have been obtained also for the Toda model with slight quenched perturbations of the potential (5).

From all these numerical experiments and other results obtained by other authors in different models (discretized version of nonlinear Klein--Gordon equation, one and two dimensional Lennard-Jones lattices, ß-FPU ecc.) one has very strong evidence that the result ε_c=const for $N \gg 1$ is model-independent (even if the value of ε_c maintain a model dependence).

The extension of the results (both analytical and numerical) obtained for hamiltonian systems with a finite (even if large) number of degrees of freedom to continuous systems, i.e. classical fields, is not straight forward. It is not useless to underline that the thermodynamic limit does not coincide with the continuous limit indeed in the first case one has $V \to \infty$ and $N \propto V$ (V is the volume) in the second one V=const and $N \to \infty$; therefore, the numerical study of the problem of equipartition is somewhat different in these two limits.

Recently Patrascioiu (6) has shown that in a large class of d-dimensional classical field theories with lagrangian density

$$[10] \quad \mathcal{L} = \frac{1}{2}\left[\dot{\varphi}^2(\underline{x},t) - (\underline{\nabla}\,\varphi(\underline{x},t))^2 - m^2\varphi^2(\underline{x},t)\right] - V(\varphi), \quad V(\varphi) \propto |\varphi|^p$$

time averages and microcanonical averages do not coincide if the initial condition has a finite energy and a global smooth solution of the equations of the motion exists. We do not repeat here Patroscioiu's argument (see ref. 7) but we want to underline that a consequence is that if the energy is initially fed in the lower modes only a very small fraction reaches the higher modes. So one obtains that the ultraviolet (u.v.) catastrophe which follows from the (Gibbs')classical statistical mechanics is incompatible with the equations of motion. Such a conclusion can be extended to the two-dimensional Euler equation:

$$[11] \quad \begin{cases} \partial_t \underline{u} + (\underline{u}\cdot\underline{\nabla})\underline{u} = -\underline{\nabla} p \\ \underline{\nabla}\cdot\underline{u} = 0 \end{cases}$$

where $\underline{u}$ is the velocity field and $-\nabla p$ is the pressure force per unit mass. Assuming, for simplicity, periodic boundary conditions on a square $(0,L)^2$ one can introduce the Fourier series represetantion of $\underline{u}$:

$$[12] \quad \underline{u}(\underline{x},t) = \sum_{|\underline{k}|<k_{MAX}} \hat{\underline{u}}(\underline{k},t)\, e^{i\underline{k}\cdot\underline{x}}, \quad \underline{k} = \frac{2\pi \underline{n}}{L}$$

Because of the conservation laws

$$[13] \quad \begin{cases} E = \frac{1}{2} \sum_{|\underline{k}|<k_{MAX}} |\hat{\underline{u}}(\underline{k},t)|^2 = \text{const.} \\ \Omega = \frac{1}{2} \sum_{|\underline{k}|<k_{MAX}} (\underline{k})^2 |\hat{\underline{u}}(\underline{k},t)|^2 = \text{const.} \end{cases}$$

and of Liouville theorem one can define the equilibrium statistical mechanics in the standard way. Introducing the canonical ensemble one obtains easily:

$$\overline{|\hat{\underline{u}}(\underline{k})|^2} \sim \frac{1}{\alpha + \beta(\underline{k})^2} \tag{14}$$

in the continuum limit ($K_{MAX} \to \infty$) [14] shows the u.v. catastrophe . It is easy to prove that the u.v. catastrophe cannot happen in the two-dimensional case starting with a smooth initial condition. Introducing the stream function :

$$\underline{u}(\underline{x}) = \underline{\nabla}^{\perp} \psi(\underline{x}) \quad , \quad \underline{\nabla}^{\perp} = (-\partial_2, \partial_1) \tag{15}$$

the continuity equation $\underline{\nabla} \cdot \underline{u} = 0$ is authomatically satisfied and eq. [11] can be rewritten as:

$$\partial_t \omega + (\underline{u} \cdot \underline{\nabla}) \omega = 0 \tag{16a}$$

$$\omega = \varepsilon_{ij} \partial_i u_j = \Delta \psi \qquad ; \; i, j = 1, 2 \tag{16b}$$

Eq. [16a] expresses the conservation of vorticity in any fluid element. From eq. [16b] one has

$$\psi(\underline{x}) = \int G(\underline{x} - \underline{x}') \, \omega(\underline{x}') \, d\underline{x}'$$

where G(x - x') is the Green function of the laplacian operator and

$$\underline{u}(\underline{x}) = \int \underline{\nabla}^{\perp}_{\underline{x}} G(\underline{x} - \underline{x}') \, \omega(\underline{x}') \, d\underline{x}' \tag{17}$$

Because of [16a] one obtains

$$\max_{\underline{x}} |\omega(\underline{x}, t)| = \max_{\underline{x}} |\omega(\underline{x}, 0)| \tag{18}$$

Besides, a classical theorem ensures

$$|\partial_i G(\underline{x})| < \frac{\text{const.}}{|\underline{x}|} \tag{19}$$

From eqs. [17], [18] and [19] it follows by straightforward computations

$$
[20] \qquad |u(\underline{x}+\underline{r}) - u(\underline{x})| < C_1 |\underline{r}| \left| \ln\left(\frac{r}{L}\right) - 1 \right| \qquad C_1 \sim \max_{\underline{x}} |\omega(\underline{x},0)| \sim \Omega^{1/2}
$$

From eq. [20] we obtain the bound for $K \gg 1/_L$:

$$
[21] \qquad |\hat{\underline{u}}(\underline{K})|^2 \leq \text{const.} \, |\underline{K}|^{-4+\epsilon}
$$

Note the eq. [21] tells us that a smooth initial condition leads to situations with a poor fraction of energy contained in the high wave number modes, in sharp disagreement with eq. [14], obtained on the basis of equilibrium statistical mechanics.

A relevant point is that Patrascioiu's results and the bound [21] seem to clash with the numerical experiments performed on hamiltonian systems with many degrees of freedom [3] (which can be considered as discretized versions of classical field theories) and on the two-dimensional Euler equation [8]. However this disagreement is only apparent; let us discuss the latter case. In the numerical experiments on the two-dimensional Euler equation (starting far from the equipartition) the system quickly approaches "thermodynamic equilibrium" (i.e. eq. [14]). For eq. [11], one can prove the existence of an analytic continuation of $\underline{u}(\underline{x}, t)$ for any t in a finite strip around the real plane of thickness $\delta(t) > 0$ (9). In any numerical simulation of a partial differential equation one must introduce a discretization of space, so that there is a lower resolution scale Δx (the length of the mesh or the inverse of the maximum wave number, if one performs a truncation in the Fourier space). It is evident that at time t such that $\delta(t) \gtrsim \Delta x$ everything works well - the numerical simulation provides a good representation of the continuum system and the equipartition of energy is absent. On the contrary when $\delta(t) \lesssim \Delta x$ the discretized system does not reproduce the analiticity properties of the continuous system. Therefore the equipartition of energy observed in numerical experiments must be considered as a feature of the discretized system. Unfortunately in two-dimensional fluid dynamics $\delta(t)$ decays exponentially in time

$$
\delta(t) \sim \exp(-at) \quad , \quad a \sim O(1)
$$

and the analyticity properties of the solution of the continuous system are rapidly lost also with very small Δx.

In order to show the role of the truncation of field equations we have studied numerically a one dimensional model of a radiant cavity (10). This model describes the elettromagnetic field between two fixed parallel mirrors (at ± 1) coupled to a charged plane, parallel to the mirrors, which is allowed to move in the direction of the mirrors. The equations of motion are

[22]
$$\partial_t^2 A - \partial_x^2 A = 2\pi^{1/2}\beta \dot{z}\, \delta(x)$$
$$\ddot{z} = -\frac{\beta}{2\pi^{1/2}} \partial_t A(0,t) - \alpha z^3$$

where $\delta(x)$ is the Dirac delta function.

Expanding A in the odd Fourier series eqs. [22] become:

[23]
$$\ddot{a}_n + \omega_n^2 a_n = \beta \dot{z} \qquad n = 1,2,\ldots \qquad \omega_n = \frac{\pi}{2}(2n-1)$$
$$\ddot{z} = -\beta \sum_{n=1}^{\infty} \dot{a}_n - \alpha z^3$$

In order to perform a numerical simulation of eqs. [23] we have truncated them, taking into account a_n with $n \in [1,N]$. Of course $N \to \infty$ corresponds to the continuum limit.

We have done two simulations at a given energy with the same initial condition and different N (N=8 and N=16). In Fig. (3) we plot the energy permode E_n vs. n:

[24]
$$E_n = \frac{1}{2}\left(\dot{a}_n^2 + \omega_n^2 a_n^2\right)$$

For N=8 the system has reached equipartition but with N=16 (we are going towards the "continuum" limit) there is a good evidence of an exponential tail, i.e. a non vanishing strip in which an analytic continuation of the field exists. In this case it is well evident how the equipartition for N=8 is an artifact of the truncation and it cannot be expected to be a property of the continuum limit.

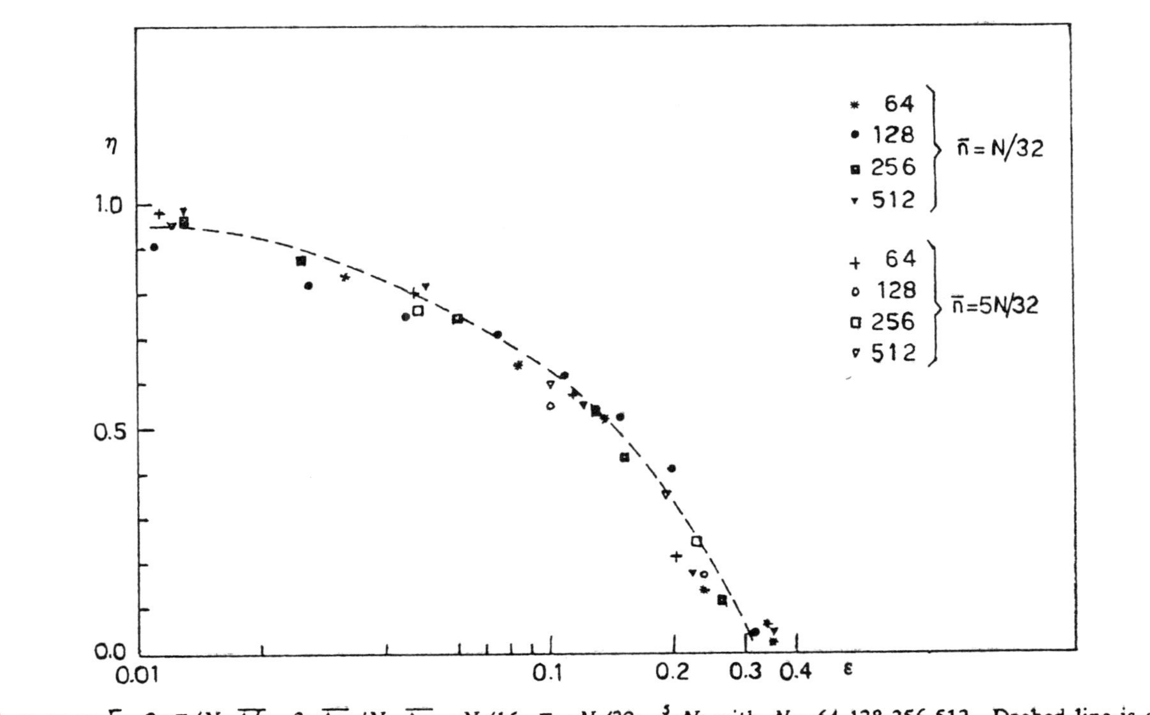

FIG. 1. η vs ϵ; $\bar{k}=2\pi\bar{n}/N$, $\overline{\Delta k}=2\pi\overline{\Delta n}/N$, $\overline{\Delta n}=N/16$, $\bar{n}=N/32$; $\frac{5}{32}N$; with $N=64,128,256,512$. Dashed line is a free-hand smoothing of the experimental results.

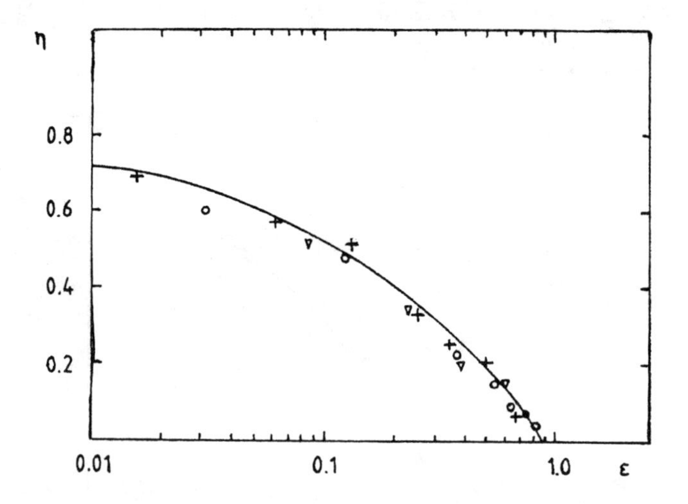

FIG. 2. Time asymptotic value of the spectral entropy vs energy density. The value of the coupling constant α has been set equal to 0.1. Circles stand for $N = 64$, $\bar{n} = 2$, $\Delta n = 4$; crosses stand for $N = 128$, $\bar{n} = 4$, $\Delta n = 8$; triangles stand for $N = 256$, $\bar{n} = 8$, $\Delta n = 16$.

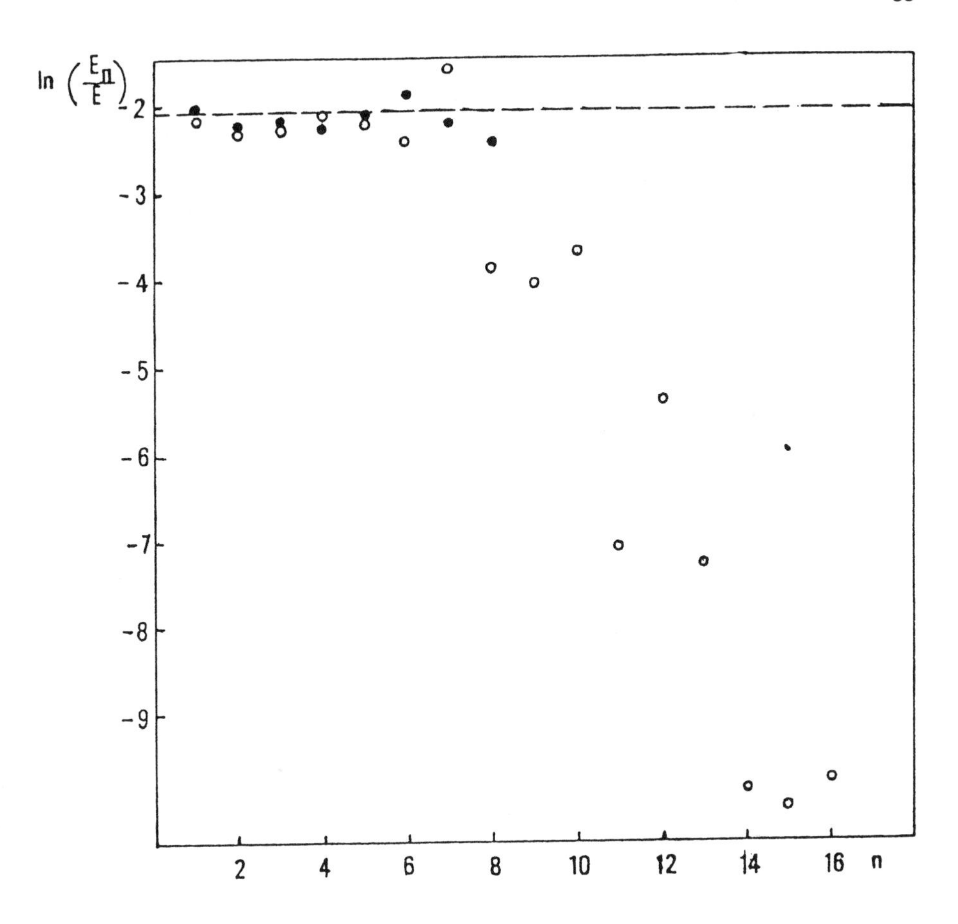

FIG. 3. Ln(E_n/E) vs. n; average is over a time interval $T=10^5$ ($\Delta t=.01$) and E is the total energy (in this case E=1297), α =1, β = 1.18. The full dots refer to the system with N=8, the circles to N=16. The initial condition in the same in both cases; the broken line indicates equipartition for N=8.

REFERENCES

(1) Fermi E., Pasta J., Ulam S.,
Los Alamos Sci. Lab. Report No LA-1940 (1955).
(2) A.N. Kolmogorov,
Dokl. Akad. Nauk. SSSR 98, 527 (1954).
Arnold V.I.,
Russ. Math. Surv. 18, 9 (1963).
Moser J.,
Nachr. Akad. Wiss. Gottingen Math. Phys. K1 2, 15 (1962).
(3) Bocchieri P., Scotti A., Bearzi B., Loinger A.,
Phys. Rev. A2, 2013 (1970).
Callegari B., Carotta M.C., Ferrario C., Lo Vecchio G. and Galgani L.,
Il Nuovo Cimento 54B, 463 (1979).
Benettin G., Lo Vecchio G. and Tenenbaum A.,
Phys. Rev. A22, 1709 (1980).
Livi R., Pettini M., Sparpaglione M., Ruffo S., Vulpiani A.,
Phys. Rev. A28, 3544 (1983).
(4) Izrailev F.M. and Chrikov B.V.,
Sov. Phys. Dokl. 11, 30 (1966).
(5) Livi R. et al., Phys. Rev. 31A, 1049 (1985).
Livi R. et al., Phys. Rev. 31A, 2740 (1985).
Isola S., Livi R., Ruffo S., Vulpiani A.,
preprint University of Florence (1985).
(6) Patrascioiu A.,
Phys. Lett. 104A, 87 (1984).
(7) Livi R., Pettini M., Ruffo S. and Vulpiani A.,
preprint University of Rome (1985).
(8) Fox D.G. and Orszag S.A.,
Phys. Fluid 16, 169 (1973).
Basdevant C. and Sadourny R.,
J. Fluid Mech. 69, 673 (1975).
(9) Sulem C., Sulem P.L., Frisch H.,
J. Comp. Phys. 50, 138 (1983).
(10) Bocchieri P., Scotti A., Loinger A.,
Lett. Nuovo Cimento 4, 341 (1972)
Benettin G., Galgani L.,
J. Stat. Phys. 27, 153 (1982).

GEOMETRY OF TRAJECTORIES AND STOCHASTICITY IN THE PHASE SPACE

Mario Casartelli
Dipartimento di Fisica dell'Università - Parma
Unità Risonanze Magnetiche del GNSM - Parma

Abstract

Two sets of observables, connected to the curvature and to the phase space velocity respectively, are introduced for numerical inspections on the invariant surfaces of hamiltonian systems. Their relations to stochasticity are discussed. Reference is made to the Hénon-Heiles system as a test model.

K.A.M.-Theory ensures us that a deep connection exists between the geometry of the phase space of a hamiltonian system and its dynamical properties. In general, such connection is however difficult to be revealed directly. Most used stochastic parameters refer indeed to the dynamical aspects, or consequences, of the order-desorder transition (decay of correlations, local divergence of orbits, trend to equipartition, etc.). And yet, an inner insight in the geometrical aspects would be extremely appreciable in order to reveal the role of the invariant surfaces, seen as local constraints, in the relaxation, especially in the low stochastic regime of motion. There exist indications on the possibility of this role, and speculations on its relevance (1). Some ideas and preliminary results on numerical studies about this problem are reported here.

In the ordered region of the 2n-dimensional phase space of a hamiltonian system, a trajectory lies on an invariant surface (a n-torus), while in the chaotic region it wanders almost everywhere in the (2n-1)-dimensional energy surface given by $H(p,q) = E$. In the passage to chaos, it is plausible that the increasing complicatedness of tori, which firstly deformate and then vanish into higher dimensional manifolds, will affect the geometrical features of orbits themselves: their curvature, for instance, or "torsion" (this last to be correctly defined). Precisely, if $x(t)$ is a parametrized curve with curvilinear coordinate $s = s(t)$, its curvature $c(t)$ is

$$c(t) = \sqrt{\sum_k \frac{d^2 x_k}{ds^2}}$$

One can compute $\langle c \rangle_T$ and $\langle s \rangle_T$ defined by

$$\langle c \rangle_T = \frac{1}{T}\int_0^T c(t)\, dt$$

$$\langle s \rangle_T = \frac{\langle c^2 \rangle_T - \langle c \rangle_T^2}{\langle c \rangle_T^2}$$

and limits as $T \to \infty$, which exist via Birkhoff Theorem, will be denoted $\langle c \rangle$ and $\langle S \rangle$ respectively. $\langle S \rangle_T$ and $\langle S \rangle$ are the relevant quantities for our purposes.

Consider now N uncoupled harmonic oscillators, whose hamiltonian H_o in suitable coordinate may be written

$$H_o = \frac{1}{2} \sum (q_k^2 + p_k^2)$$

Every orbit of this integrable system has constant curvature $c(t) = \sqrt{1/2E}$, E being the energy, independently from the initial conditions. Therefore, $\langle c \rangle_T = \langle c \rangle = \sqrt{1/2E}$, $\langle S \rangle_T = \langle S \rangle = 0$. A weak coupling term $H'(p,q)$ destroys analytical integrability, but not the most part (in terms of measure) of the invariant surfaces, which simply will result deformate. Small changes in curvature will compensate each other, so that $\langle c \rangle_T$ cannot be used as an index of the incoming disorder. Summing up relative deviations along the trajectory, $\langle S \rangle_T$ can give on the contrary a quantitative estimate of the deformations of the tori, both locally (i.e. following the behaviour in T) and globally (i.e. in the limit as $T \to \infty$). For a strong coupling, the fluctuation of the curvature is expected to grow in such a way to be directly detectable from $\langle S \rangle_T$. Consider, in particular, a low stochastic situation: the energy surface is metrically transitive, and yet there exist invariant surfaces (even if their measure has gone to 0) which can affect for finite times the local behaviour of the trajectories. There is likelihood that such influence shall appear from the T dependence of $\langle S \rangle_T$.

Similar considerations may be developed for another set of quantities $\rho(t)$, $\langle \rho \rangle_T$, $\langle G \rangle_T$, $\langle \rho \rangle$ and $\langle G \rangle$ (defined as $c(t)$, $\langle c \rangle_T$, $\langle S \rangle_T$, $\langle c \rangle$ and $\langle S \rangle$ before) starting from

$$\rho(t) = \left| 1/\nabla H \right|$$

The Hamilton equations give

$$\rho(t) = |v_P|^{-1}$$

where $v_P = (\dot{p},\dot{q})$ denotes the phase space velocity. The natural relation between time of sojourn and probability justifies, for ergodic systems, the interpretation of $\rho(t)$ as Gibbs ensemble, the microcanonical density indeed. Nevertheless, $\rho(t)$ and related quantities are meaningful phase observables independently from ergodicity. (Note also the independence of $\langle G \rangle_T$ from any normalization factor). The procedure outlined above may be repeated starting from $\rho(t)$ and from the observation that, once again, for uncoupled oscillators, $\rho(t) = \sqrt{1/2E}$ = const. The numerical analysis presents simpler in this case, since the computation of second derivatives along the orbit is no more required. Curvature deserves however to be considered for its intrinsic geometrical meaning, and moreover simultaneous experiments on the two sets of observables will sustain each other in unclear situations.

The criteria exposed above have to be checked on a system whose ergodic and geometrical features are fairly well known through independent analysis. The Hénon-Heiles model, i.e. the dynamical system with hamiltonian

$$H(p,q) = \frac{1}{2}(p_1^2 + p_2^2 + q_1^2 + q_2^2) + q_1^2 q_2 - q_2^3/3$$

is a natural candidate: the fact that its phase space is only four dimensional gives indeed a particular transparency to the classical pictures of its surfaces of section and to their connections with dynamical properties (2). A counter indication consists in the existence of an escape energy, which stops the analysis at a relatively low stochastic level. A qualitative pattern of the results of computations performed on this system, which will be exposed in details elsewhere (3), is the following:

1 - Indications from curvature and "microcanonical density" are always coherent and compatible. Generally speaking, not only the second set of observables is easier to compute, as expected, but it gives also neater results.

2 - By considering $\langle S \rangle$ and $\langle G \rangle$ as functions of E, the stochastic transition of the model is refound at the same values obtained through other methods: this makes $\langle S \rangle$ and $\langle G \rangle$ possible alternative stochastic parameters.

3 - The relaxation in the low stochastic regime of motion, observed through the T-dependence of $\langle S \rangle_T$ and $\langle G \rangle_T$, shows a clear dependence on initial conditions, giving evidence to the constraining action of the invariant surfaces. In this context, the Cesaro resummation of the time series proves useful to extrapolate the limit values we spoke about in point 2.

In conclusion, the experiments on the Hénon-Heiles model confirm the expected qualitative properties of $\langle S \rangle$ and $\langle G \rangle$, and authorize in perspective their use in the analysis of other and more complicate systems.

REFERENCES

(1) M.Casartelli: Nuovo.Cim. 76B, 97 (1983); idem, in Dynamical Systems and Chaos, Proc. Sitges 1982, L.Garrido ed. (N.Y. Heidelber Berlin 1983); G.Benettin and L.Galgani: J.Stat.Phys. 27, 153 (1982); see also L.Galgani, this conference.

(2) M.Hénon and C.Heiles: Astron.J. 69, 73 (1964); F.Gustavson: Astron. J. 21, 670 (1966); for general references, A.J.Lichtenberg and M.A. Liebermann: Regular and Stochastic Motion (N.Y. Heidelberg Berlin 1983)

(3) M.Casartelli and S.Sello, in preparation.

CANONICAL PERTURBATION SERIES AND BREAK UP OF INVARIANT TORI

G. Turchetti

Dipartimento di Fisica dell'Università di Bologna and
Istituto Nazionale di Fisica Nucleare, Sezione di Bologna,
Via Irnerio n. 46, 40126 Bologna, Italy.

ABSTRACT

The perturbative solution of the conjugation problem for Hamiltonian systems is considered. For area preserving maps the conjugation in a region of phase space produces asymptotic (Birkhoff) series while the conjugation of a single invariant curve with a circle, leads to series whose convergence seems to persist up to the break up point. The origin of convergence and divergence is found to be in the different way the resonant contributions in the divisors accumulate.

The analysis of the Siegel problem confirms such a picture.

1. Introduction

When a simple dynamical system is perturbed in such a way that the topological structure of its orbits is not varied, one can still conjugate it with a simple system.

Among the hamiltonian systems the integrable ones are simple, since their orbits belong to the tori foliating the energy surface. In this case the conjugation of the perturbed system with an integrable one cannot be achieved in an open set of phase space since any perturbation modifies the topological structure of some orbits [1]. However even though a dense set of tori is destroyed the surviving ones can be smoothly interpolated [2]. As a consequence for small enough perturbations the destroyed tori escape numerical observations and the formal series which solve the conjugation equations are asymptotic to the interpolated tori; with suitable truncations these series provide extremely accurate approximations [3].

By investigating the simpler models of area preserving maps a mechanism of accumulation of resonant contributions was found to be responsible for the asymptotic character of these perturbative (Birkhoff) series [4].

For large perturbations the destruction of tori becomes macroscopically evident and the Birkhoff series start to exhibit a divergent behaviour. In order to describe this transition to chaos one would like to compute the critical value of the perturbation strenght λ which cause the break up of a single torus and to understand the break up mechanism. Once more the perturbation methods prove to be useful: a convergent and computable perturbative expansion is obtained by tracking the torus, i.e. by keeping its frequencies, fixed (at rationally independent diophantine values), while the perturbation strenght λ is varied. The perturbed hamiltonian is no longer conjugated with an integrable hamiltonian but only the transformation of a distorted torus with a plain torus is considered.

The conjugation becomes a local transformation in phase space and with suitable analyticity hypothesis on the hamiltonian is represented by convergent series. The analysis of simple model such as the mapping of the circle [5], the standard map [6] and the Siegel problem [7] has shown that the series do converge up to the break up of the torus. This occurs since the analyticity domain in the torus coordinate is reduced by increasing λ and becomes just the real line at a critical value λ_c. Differentiability is also lost and at λ_c the torus becomes a fractal. The same phenomenon is observed in the Siegel problem were the boundary of the Siegel domain, which is filled with invariant curves, is a fractal, image of the convergence circle, which is also natural boundary of the non conjugation function.

For the hamiltonian cases with two or more degrees of freedom one can obviously expect a similar picture and, even though rigorous proof are not likely to be obtained, numerical evidence could certainly be provided.

2. Area preserving maps.

To the flow of a hamiltonian system with two degrees of freedom can be associated an area preserving map using a Poincaré section. Since the map gives all the relevant information on the flow itself it is customary to study model maps without explicit reference to the hamiltonian flow. We consider two types of maps obtained by perturbing an integrable map which has an elliptic fixed point at the origin: syncronous if the unperturbed map has a fixed rotation angle

$$\begin{cases} x' = [x + F(x,y)]\cos\alpha - [y + G(x,y)]\sin\alpha \\ y' = [x + F(x,y)]\sin\alpha + [y + G(x,y)]\cos\alpha \end{cases} \tag{2.1}$$

and asyncronous if the rotation angle of the unperturbed map varies

$$(2.2) \quad \begin{cases} r' = r + \lambda f(r,\theta) \\ \theta' = \theta + r + \lambda g(r,\theta) \end{cases}$$

In (2.1) we require that F.G vanished at the origin with their first derivatives and that $\frac{\partial(x',y')}{\partial(x,y)} = 1$. In (2.2) f, g are periodic functions of θ with period 2π and such that $\frac{\partial(r',\theta')}{\partial(r,\theta)} = 1$. We remark that any asyncronous map could be represented by 2.1 where α is a function of $x^2 + y^2$ and any syncronous one by 2.2 where the unperturbed rotation angle r is replaced by a constant α . The form (2.2) includes all the asyncronous maps with monotic rotation angle $\alpha(r)$ if one scales the radial coordinate. The area preserving transformation from (x,y) to (r, θ) is obviously $x = \sqrt{2r}\cos\theta$, $y = \sqrt{2r}\sin\theta$.

3. The Birkhoff series

When $\alpha/2\pi$ is irrational the orbits of (2.1) are dense on circles if F = G = 0. When F, G$\neq$0 if the circles were only distorted one could conjugate $\mathcal{J}$ with and integrable map $\mathcal{K}$

$$(3.1) \quad K \begin{cases} \xi' = \xi \cos\Omega(\xi^2+\eta^2) - \eta \sin\Omega(\xi^2+\eta^2) \\ \eta' = \eta \sin\Omega(\xi^2+\eta^2) + \eta \cos\Omega(\xi^2+\eta^2) \end{cases}$$

via a transformation $T : (\xi, \eta) \rightarrow (x,y)$

$$(3.2) \quad T \begin{cases} x = \xi + u(\xi,\eta) \\ y = \eta + v(\xi,\eta) \end{cases}$$

$$(3.3) \quad \mathcal{J} \circ T = T \circ \mathcal{K}$$

A formal solution of (3.3) as power series in ξ, η was given by Birkhoff choosing the distance from the origin as perturbation parameters and reads

$$(3.4) \quad u = \sum_{n\geq 2} u_n(\xi,\eta);\ v = \sum_{n\geq 2} v_n(\xi,\eta)$$

where u_n, v_n are homogeneous polynomials of order n recursively determined together with the coefficients Ω_{2m} of the series $\Omega(\xi^2+\eta^2) = \alpha + (\xi^2+\eta^2)\cdot\Omega_2 + \ldots$, see reference [8]

The series (3.4) are asymptotic to the transformation $\tilde{T}$ which interpolates the existing invariant curves and can be conveniently analysed writing it as a Fourier series. Letting $z = x + iy$, $\xi = \xi + i\eta = \sqrt{\rho}\, e^{i\tau}$ we have:

$$(3.5) \qquad z = T(\xi) = \sum_{K=-\infty}^{+\infty} T_K(\rho)$$

where

$$(3.6) \qquad T_K(\rho) = \rho^{|K|/2} \sum_{n=0}^{\infty} \rho^n a_K^{(n)}$$

The arguments which follow are all based on a thorough analysis of the quadratic map [9] $F=0$, $G=-x^2$ whose phenomenological description can be found in [10]. If we truncate the series for T at order N in the variable $\sqrt{\rho}$ so that only terms with $2n + |K| \leq N$ survive in (3.6), then (3.3) is satisfied with an error $\mathcal{O}(|\xi|^N)$ as first noticed in [11].

The divergence when $N \to \infty$ appears in the Fourier coefficients themselves and is due to the accumulation of resonant contributions. Choosing for $\alpha/2\pi$ a quadratic irrational we label with j_s/N_s its continued fraction approximations and write

$$(3.7) \qquad \alpha = 2\pi \frac{j_s}{N_s} + \varepsilon_s \qquad |\varepsilon_s| \leq \frac{1}{N_s^2}$$

When $K = N_s + 1$ or $K = -N_s + 1$, the Fourier coefficient T_K first receives the contribution ε_s^{-1} from the resonance N_s in the divisor $(e^{-i\alpha N_s} - 1)$ and all the subsequent terms keep on receiving ε_s^{-1} contributions so that

$$(3.8) \qquad a_K^{(n)} \simeq \varepsilon_s^{-n}$$

The precise relation reads

$$a^{(n+1)}_{-N_s+1} = -\frac{N_s+n+1}{N_s+n}\,\frac{\Omega_2}{\varepsilon_s}\,a^{(n)}_{-N_s+1} + O(1)$$

(3.9)

$$a^{(n+1)}_{N_s+1} = -\frac{n+2}{n+1}\,\frac{\Omega_2}{\varepsilon_s}\,a^{(n)}_{N_s+1} + O(1)$$

where $\Omega_2 = -\frac{1}{16}\left[3\cot g\frac{\alpha}{2} + \cot g\frac{3}{2}\alpha\right]$ is the first coefficient in the expansion of Ω. As a consequence the series for T_{-N_s+1} and T_{N_s+1} are geometric with ratios $\rho_s \sim |\varepsilon_s/\Omega_2|$; if for N sufficiently larger than N_s we increase ρ beyond ρ_s, these Fourier coefficients will increase and the image by T of a circle will first exhibit N_s cusps and then N_s loops (see fig. 1) whose size increases with ρ. If one considers the images of a circle by $z = e^{i\tau} + \lambda e^{i(N+1)\tau}$ it is easy to show that the N cusps arise when $\lambda = \frac{1}{2(N+1)}$.

For $N > N_s$ and ρ larger than ρ_s the series does no longer define a one to one transformation. If we keep ρ below ρ_s but increase N beyond N_{s+1} the next resonance will act. The Fourier coefficients with $K = -N_{s+1}+1,\ N_{s+1}+1$ will start as geometric series with ratio $\sim|\varepsilon_{s+1}/\Omega_2|$; the lower Fourier coefficients T_k will also exhibit a similar behaviour, as one can see in figure 2, after some delay due to the time required for this resonance to prevail on the previous resonant contribution. By increasing ρ beyond $\rho_{s+1} = |\varepsilon_{s+1}/\Omega_2|$ for $N > N_{s+1}$ the rise of N_{s+1} loops will be observed, because the Fourier coefficient A_k with $K = -N_{s+1}+1,\ N_{s+1}+1$ become important.

The structure of any Fourier coefficient is dominated by the infinite sequence of resonances N_s and the ratios of coefficients have constant values $\rho_s = |\varepsilon_s/\Omega_2|$ on intervals of length $\approx N_{s+1} - N_s$. Since $\rho_s \to 0$ as $s \to \infty$ the final radius of convergence is actually zero.

4. Evolution and break up of an invariant curve

The evolution with λ of an invariant curve with fixed winding number $\frac{\omega}{2\pi}$ (ω is defined as the average rotation angle) satisfying the diophantine condition $\left(|\omega/2\pi - m/n|^{-1} \leq c n^{\mu},\ \forall m, n \in \mathbb{Z}\right)$ was analysed in [6] for the standard map defined by (2.2) with f = g = sin θ.

An invariant curve with winding number $\omega/2\pi$ can be conjugated with a circle, on which the dynamics is a simple translation $\mathcal{K}$

$$\mathcal{K}\begin{cases}\omega' = \omega \\ \tau' = \tau + \omega\end{cases} \tag{4.1}$$

by the transformation

$$T\begin{cases} r = \omega + \lambda v(\tau,\omega;\lambda) \\ \theta = \tau + \lambda u(\tau,\omega;\lambda)\end{cases} \tag{4.2}$$

The functional equation for the conjugation is still given by (3.3) and explicitly reads

$$\begin{cases} u(\tau+\omega) - 2u(\tau) + u(\tau-\omega) = \lambda \sin(\tau + u(\tau)) \\ v(\tau) = u(\tau) + u(\tau-\omega)\end{cases} \tag{4.3}$$

omitting for brevity the dependance on ω and λ .

Since the functional equation (3.3) has an analytic solution in λ and τ [12], the behaviour of the invariant curve was analysed after obtaining a series solution of (4.3) according to

$$u(\tau) = \sum_{K=-\infty}^{+\infty} A_K(\lambda)\, e^{ik\tau} \tag{4.4}$$

with

$$A_K(\lambda) = \lambda^{|K|} \sum_{n=0}^{\infty} \lambda^n a_K^{(n)} \tag{4.5}$$

The series for the Fourier coefficients $A_K(\lambda)$ have a convergent behaviour : indeed even though $|a_K^{(n)}|^{-1/n}$ exhibit plateaus at $2(N_s - |K|)$ (see figure 3), the jump from one plateau to the next decreases sufficiently rapidly to allow convergence. It has been shown [9] that even though the recursion defining the coefficients $a_K^{(n)}$ is very similar to the one defining the Birkhoff series, cancellations occur such that the resonance which first gives a contribution ε_s^{-2} to the coefficient $a_{N_s}^{(0)}$ does not keep on contributing to the $a_{N_s}^{(n)}$ for $n>0$. As a consequence $a_{N_s}^{(n)} \simeq \varepsilon_s^{-2}$ for $0 \le n \le 2(N_{s+1}-N_s)$ so that the coefficients radically change only when the next resonance is encountered. With such a mechanism a convergent series is built.

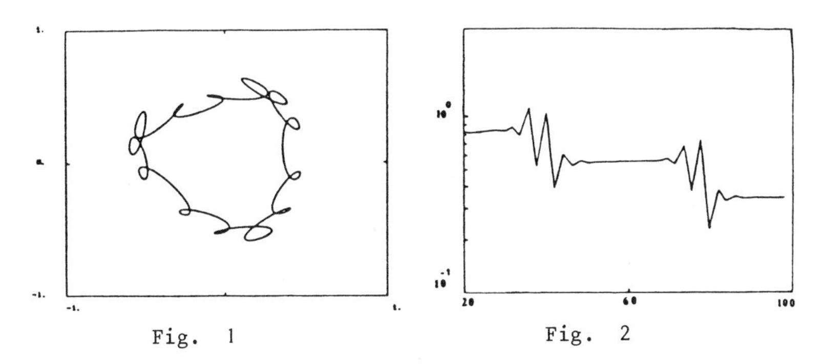

Fig. 1 Fig. 2

Fig. 1 - Curve with 8 loops which shows for the Birkhoff series of the quadratic map with $\omega = (\sqrt{5} - 1)/2$ the divergence due to the resonance $N_4 = 8$.

Fig. 2 - Sequence of ratios $|a^{(n)}/a^{(n+1)}|$ of the first Fourier coefficient $A_1(\rho)$ for the quadratic map.

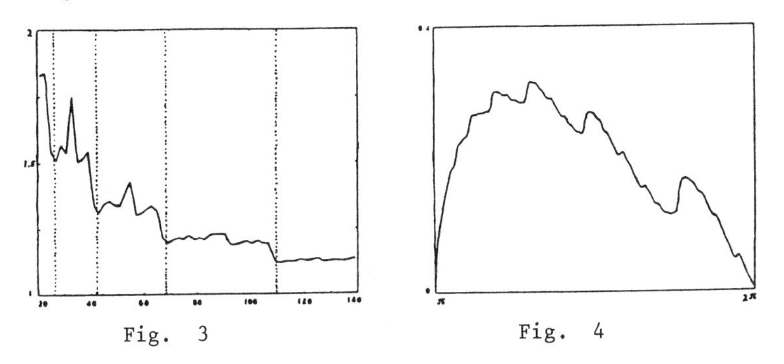

Fig. 3 Fig. 4

Fig. 3 - Sequence $|a^{(n)}|^{-1/n}$ for the standard map with $\omega = (\sqrt{5} - 1)/2$. The dotted lines are at $2N_s$, where N_s are the Fibonacci numbers.

Fig. 4 - Plot of the critical curve $u(\tau, \lambda_c)$ for the Standard map.

Choosing a winding number equal to the golden mean $\frac{\omega}{2\pi} = \frac{\sqrt{5}-1}{2}$ the radius of convergence, up to the truncation order N = 140, was found larger than 1, (see figure 2), so that $A_k(\lambda)$ for $|\lambda|<1$ could be computed without significant errors [6]. The Fourier spectrum $A_k(\lambda)$ decreases exponentially as $e^{-\beta(\lambda)|k|}$; $\beta(\lambda)$ decreases with λ and vanishes for $\lambda = \lambda_c = .971$. At this critical value, which agrees with Greene's value [13], a subsequence $A_{N_s}(\lambda_c)$ of Fourier coefficients N_s is found which decreases linearly $|A_{N_s}(\lambda_c)| \simeq \frac{1}{N_s}$ and $u(\tau,\lambda_c)$ is continuous but no longer differentiable, (see, figure 4). Indeed a nomalization group analysis, shows that it is a fractal with self similar properties [14, 15].

In reference [5] the map of the circle onto itself was studied. The map defined by

(4.6) $$\theta' = \theta + \omega + \delta(\lambda) + \lambda \sin\theta$$

is conjugated to translations

(4.7) $$\tau' = \tau + \omega$$

by the transformation

(4.8) $$\theta = \tau + \lambda u(\tau, \lambda)$$

Due to the phase locking [16] (the winding number has plateaus at rational values) a counterterm $\delta(\lambda)$ is added to keep the winding number fixed to $\omega/2\pi$ and allows an analytic conjugation. Basically the same results as for the standard map are obtained for the convergence of the series, the behaviour of the Fourier spectrum (see figure 5), and the structure of the critical curve (see figure 6). This curve is still a fractal, as shown in [17] by using a different approach.

5. The Siegel Problem

We consider an analytic map $\mathcal{F}$ (no longer area preserving) defined by

(5.1) $$z' = e^{i\omega} z + f(z)$$

where $f(z) = O(z^2)$. For $f=0$ the circles $|z|$ = cost. are invariant curves for the map. If $f(z)$ is non zero and analytic and $\frac{\omega}{2\pi}$ satisfies the diophantine condition, then the map has invariant curves analytically diffeomorphic to circles [18]. The conjugation with

(5.2) $$\xi' = e^{i\omega}\xi$$

is obtained by an analytic transformation T defined by

(5.3) $$z = \xi + u(\xi)$$

where $u(\xi) = O(\xi^2)$. The functional equation (3.3) then reads

(5.4) $$u(e^{i\omega}\xi) - e^{i\omega}u(\xi) = [\xi + u(\xi)]^2$$

and its perturbative solution

(5.5) $$u = \sum_{n=2}^{\infty} u_n \xi^n$$

is determined by a recursion, which, for the quadratic map $f(z)=z^2$, explicitly reads

(5.6) $$u_n = (e^{in\omega} - e^{i\omega})\left(\delta_{n,2} + 2u_{n-1} + \sum_{k=2}^{n-2} u_k u_{n-k}\right)$$

If we let $\xi = \lambda e^{iz}$, we realize that (5.5) is analog to (4.4); the Siegel problem is simpler, since the Fourier coefficients A_k are given by a single term $\lambda^k u_k$ rather than a series.

The convergence of (5.5) still occurs because the u_k receive a resonant contribution ε_s^{-1} only at $n = N_s$. In figure 5 the $|u_n|^{-\frac{1}{n}}$ for $\omega = \frac{\sqrt{5}-1}{2}$ are shown; the approach to the convergence radius $\lambda_c = .326$ is slow and occurs with plateaus starting at the resonances N_s. On the circle $|\xi| = \lambda_c$ the series seems to be summable to a continuous non differentiable function. The image of the circle $|\xi| = \lambda_c$ is the trajectory of the critical point, which gives the boundary of the Siegel domain, limiting all the invariant curves of the map. This critical curve is a fractal [19] and the disc $|\xi| = \lambda_c$ is the natural boundary of the conjugation function $u(\xi)$.

6. The hamiltonian problem

In principle the analysis of the break up of tori using perturbation series can be applied to hamiltonian systems. Given the hamiltonian

(6.1) $$H = H_0(j) + \lambda V(j,\varphi) \quad ; j \in \mathbb{R}^d, \varphi \in \mathbb{T}^d$$

and letting $\Omega_0(j) = \frac{\partial H_0}{\partial j}$, $A_0(j) = \frac{\partial^2 H_0}{\partial j \partial j}$ we choose $j = J_0$, such that $\Omega_0(J_0) = \omega$ where $|\omega \cdot k|^{-1} \leq \|k\|^r$ for $k \in \mathbb{Z}^d$ and $A_0(J_0)$ is an invertible matrix. Suppose we have computed the series defining

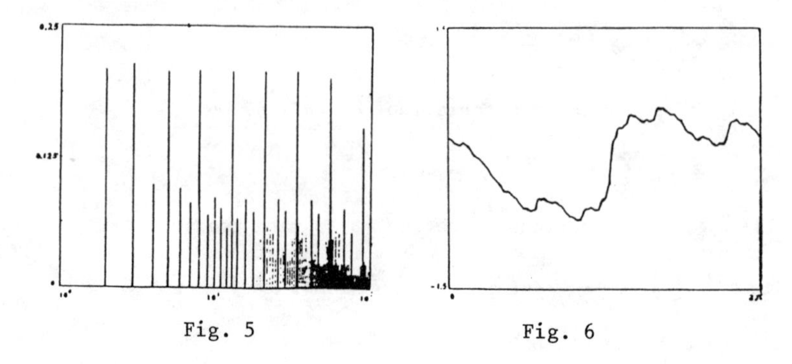

Fig. 5 Fig. 6

Fig. 5 - Fourier spectrum for the map of the circle at the critical value $\lambda_c = 1$.

Fig. 6 - Plot of the critical curve $u(\tau, \lambda_c)$ for the map of the circle.

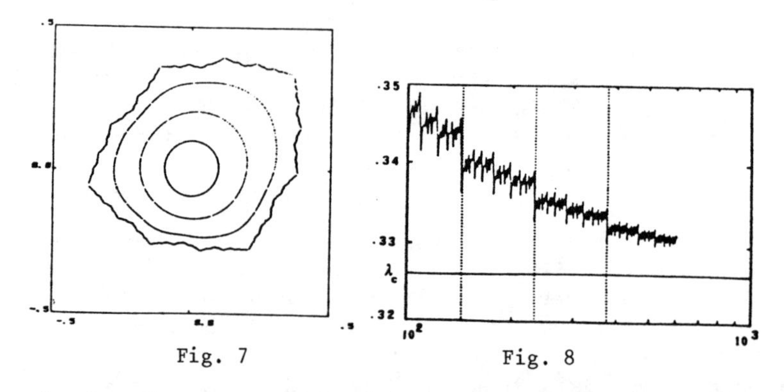

Fig. 7 Fig. 8

Fig. 7 - Invariant curves and critical curve for the Siegel problem.

Fig. 8 - Sequence $|u|^{-1/n}$ for the Siegel problem with $f(z) = z^2$ and $\omega = (\sqrt{5} - 1)/2$. The dotted lines are at the Fibonacci numbers N_s.

the transformation $(j,\varphi)\to(J,\phi)$ by solving the Hamilton-Jacobi équation

(6.2) $$H\left(\varphi, \frac{\partial F}{\partial \varphi}(\varphi, J)\right) = K(J)$$

where F denotes the generating function of the transformation and K the new integrable hamiltonian. We fix the frequency of the system to ω by inverting

(6.3) $$\Omega(J;\lambda) = \omega$$

and obtain $J = J(\omega,\lambda)$ as a series in λ

(6.4) $$J = \sum_{n=0}^{\infty} \lambda^n J_n$$

where, for instance,

(6.5) $$\Omega(J_0) = \omega \quad , \quad J_1 = -A_0^{-1}(J_0)\,\Omega_1(J_0)$$

We can then find the trasformation from ϕ to φ by replacing (6.4) in

(6.6) $$\phi = \varphi + \sum_{n=1}^{\infty} \lambda^n \frac{\partial F_n(\varphi, J)}{\partial J}$$

to obtain

(6.7) $$\phi = \varphi + \sum_{n=1}^{\infty} \lambda^n u_n(\varphi, J_0)$$

where, for instance,

(6.8) $$u_1 = \frac{\partial F_1(\varphi, J_0)}{\partial J_0} \quad , \quad u_2 = \frac{\partial F_2(\varphi, J_0)}{\partial J_0} + \frac{\partial^2 F_1(\varphi_1, J_0)}{\partial J_0 \partial J_0} J_1$$

The u_n are periodic functions in φ and are affected by divisors $(\omega\cdot k)^{-s}$. If H is analytic, then for $|\lambda|$ small enough the series will converge and the transformation $\Phi \to \varphi$ must be analytic in a strip $|\mathrm{Im}\,\Phi_s| \le \alpha_s$ [1].

We expect that at least for simple models like coupled rotators

$$H = \sum_{k=1}^{d} \frac{\dot{j}_k^2}{2} + \lambda V(\varphi_1, \ldots, \varphi_d) \qquad (6.9)$$

where V is a trigonometric polynomial, one can compute the Fourier spectra and show that they decrease exponentially up to the break point, just as for the standard map.

Conclusions

In this note previous work performed in collaboration with G. Benettin, G. Servizi and G. Zanetti [4,5,6,9] has been reviewed. The analysis of various models strongly suggests that the perturbation series which solve the conjugacy equations are adequate to give a quantitative description of the dynamics in the quasi integrable region and of the break up features of the invariant tori. The basic mechanism which makes the Birkhoff series diverge and the series for the conjugation of a single invariant curve with a circle converge, are at least partially understood: proofs and an exhaustive description of this mechanism can be found in [9].

References

(1) A.N. Kolmogorov,
Akad. Nauk. S.S.S.R., Doklady 98, 527 (1954);
V. Arnold,
Russ. Math. Surv. 18, 85 (1963).

(2) J. Chierchia, G. Gallavotti,
Nuovo Cimento 67B, 277 (1982).

(3) F. Gustavson,
Smithsonian Astrof. Obs. 90 (1962).

(4) G. Servizi, G.Turchetti, G. Benettin, A. Giorgilli,
Phys. Lett. 95A, 11 (1983).

(5) G. Zanetti, G. Turchetti,
Lett. Nuovo Cimento 41, 90 (1984).

(6) G. Benettin, G. Turchetti, G. Zanetti,
Phys. Lett. 105A, 436 (1984).

(7) G. Liverani, G. Servizi, G. Turchetti,
Lett. Nuovo Cimento 39, 417 (1984).

(8) C.L. Siegel, J. K. Moser,
"Lectures in Celestial Mechanism", Springer Verlag, Berlin, 155 (1971).

(9) G. Servizi, G. Turchetti, G. Zanetti,
"Perturbative expansions for area preserving maps",
Dip.to Fisica Bologna preprint (1985).

(10) M. Henon,
Quarterly Appl. Math. 27, 231 (1963).

(11) J. Roels M. Henon,
Bull. Astr. Soc. 2, 267 (1967).

(12) J. Moser,
Nacr. Akad. Wiss. Gottingen Math. Phys. KL IIa, 1, 1 (1962).

(13) J. Greene
J. Math. Phys. 20, 1183 (1979).

(14) S.J. Shenker, L.P. Kadanoff,
J. Stat. Phys. 27, 631 (1982).

(15) R.S. Mc Kay,
Physica 7D, 283 (1983).

(16) V.I. Arnold,
Trans. Am. Math. Soc. II Series 46, 213 (1965).

(17) S.J. Shenker,
Physics D5, 405 (1982).

(18) As in reference |8| pag 183.

(19) N.S. Manton, M. Nauenberg,
Comm. Math. Phys. 89, 555 (1983).

A MODEL FOR THE BEAM-BEAM INTERACTION AS AN EXAMPLE OF NON-INTEGRABLE HAMILTONIAN SYSTEM

G. Barbagli

Dipartimento di Fisica, Università degli Studi di Firenze,
Largo E. Fermi n. 2, I-50125 Firenze, Italy

ABSTRACT

A model of the beam-beam interaction for unbunched proton beams (weak strong case) as a map of the Poincaré surface of section is considered; the features of phase trajectories are discussed; two classical perturbation methods (Birkhoff series and Dragt transfer maps) are examined and used to derive the approximate shape of the orbits.

1. Non-integrable systems.

It is well known that only a few Hamiltonian systems are integrable, namely it is possible to define at each point of their phase space a set of 2n action-angle variables $(\underline{I}, \underline{\varphi})$, satisfying the simple equations of motion:

$$\begin{cases} \dot{\underline{\varphi}} = \underline{\omega}(\underline{I}) \\ \dot{\underline{I}} = 0 \end{cases} \tag{1}$$

The phase space of an integrable system is filled with n-dimensional invariant tori (labelled by the n action variables) on which the motion is generally quasi-periodic. The motion is ergodic if all the frequencies ω_i are not commensurable; it is periodic if they are all commensurable. The integrability of a given system is strictly related to the existence of integrals of motion, whose presence reduces the dimension of the manifold where the orbits are constrained to lie.

Within the classically known integrable problems we can quote:

1) all the problems with 1 degree of freedom and H (q,p) analytical function;
2) all the linear systems;
3) motion of a particle in a central field (and two-body problem with central forces;
4) Eulerian and Lagrangian motion of a rigid body;
5) some special problems, only solvable in the Hamilton-Jacobi formula-

tion;

6) some 1-dimensional oscillator chains.

The non-integrable systems have a much more complex structure in the phase space. They can be treated by perturbation theory, which, in its original version, is plagued by two difficulties: the secular terms (eliminated by the method of averaging) and the small denominators. An important result concerning nearly-integrable systems is the K.A.M. theorem. Through the bounds on the small denominators which hold in a region of full measure and the application of superconvergent methods one succeeds in showing that, assuming analyticity of the unperturbed Hamiltonian H $(\underline{I})$, analyticity and multiperiodicity of the perturbation Hamiltonian $H_1(\underline{I}, \underline{\varphi})$ and nondegeneracy, most of the invariant tori survive to perturbation. The theorem excludes the stochastic diffusion (Arnold diffusion) for n=2 (under the assumption of isoenergetic nondegeneracy).

2. Method of the surface of section.

Let us consider an integrable system with n=2. The motion corresponds to an area preserving map T of the Poincaré plane of section into itself. T is a rotation of an angle which depends on the radius. The circles are invariant curves.

By perturbing the system we obtain a new map T, for which some invariant curves (the ones that have a sufficiently irrational ratio of the rotation angle and 2π) survive; the remaining ones are destroyed, leaving an even number of fixed points, half elliptic, half hyperbolic, as stated by Poincaré geometric theorem.

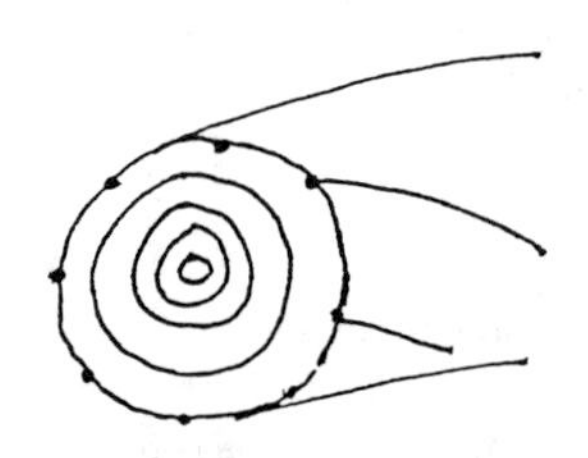

Fig. 1

Surface of section invariant curves for an integrable system.

The fixed points of the power maps T^m, T_ε^m correpond to periodic orbits. The elliptic fixed points are surrounded by stability islands; the hyperbolic ones are surrounded by instability islands, which are the sources of the large scale chaotic behaviour, shown by numerical experiments.

Definitively the relevant features of the non-integrable behaviour are [1] :

- survival of some invariant curves;
- appearance of omoclinic points;
- Arnold diffusion (proved only for particular systems, but to be expected for generic systems with $n \geq 3$).

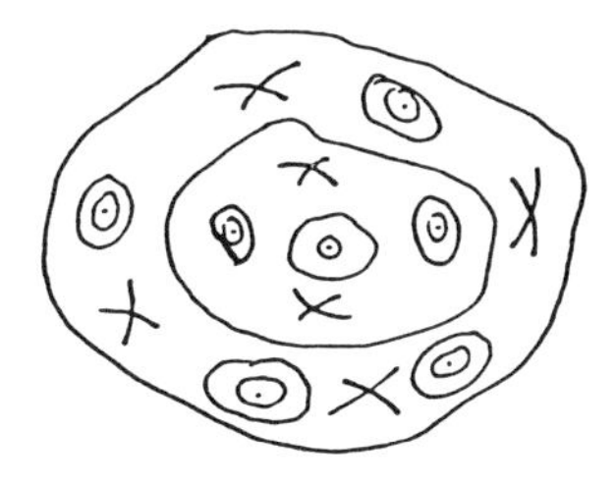

Surface of section of a perturbed integrable system.

3. Beam-beam interaction in storage rings.

A storage ring roughly consists of a vacuum chamber, where the particles circulate, and optics elements, which drive them in their motion: bending dipole magnets, providing the vertical guide magnetic field, quadrupoles focusing the transverse (betatron) oscillations, sextupoles etc.. We have also RF cavities to accelerate or to counterbalance the energy losses.

Let us introduce a reference frame joint to the ideal particle, following a circular orbit without transverse oscillations. Betatron oscillations are approximately ruled by the equation:

(2) $$x'' + k(s)x = 0$$

x is x or y; s is the coordinate along the reference orbit; k(s) is the rigidity; the differentiation is respect to s.
These are pseudo-harmonic oscillations; they can be treated using Floquet theory.

The result of moving across an element of optics whose lenght is L is given in terms of the twist matrix M:

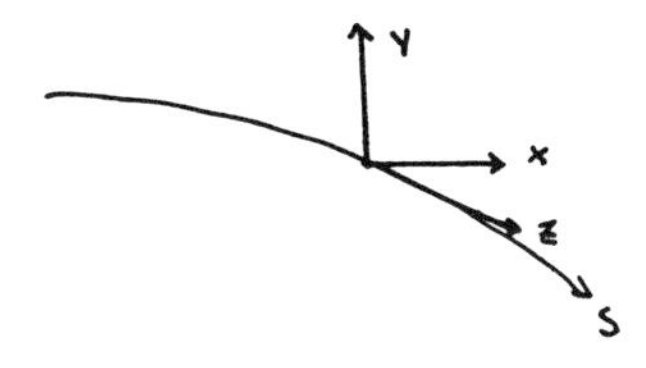

Fig. 3

Ideal reference frame.

$$(3)\quad \begin{pmatrix} x \\ x' \end{pmatrix}_{n+1} = \begin{pmatrix} \cos\mu + \alpha\sin\mu & \beta\sin\mu \\ -\gamma\sin\mu & \cos\mu - \alpha\sin\mu \end{pmatrix} \begin{pmatrix} x \\ x' \end{pmatrix}_n$$

$$(4)\quad \alpha = -\beta'/2$$

$$(5)\quad \gamma = \frac{1+\alpha^2}{\beta}$$

$$(6)\quad \mu = \int_s^{s+L} \frac{ds}{\beta(s)} = \nu \, \frac{2\pi}{N}$$

N is the number of identical sectors into which the ring is divided; ν is the betatron tune (in the following we call ν the effective betatron tune, which is 1/N of the global one).
The transformation leaves invariant the quantity (emittance)

$$(7)\quad \varepsilon = \gamma x^2 + 2\alpha x x' + \beta x'^2$$

Its level lines are ellipses in the xx' plane.

Changing variables

$$(8)\quad \begin{cases} Q = x/\beta^{1/2} \\ P = (\alpha x + \beta x')/\beta^{1/2} \end{cases}$$

the map becomes a rotation of an angle

$$(9)\quad \begin{pmatrix} Q \\ P \end{pmatrix}_{n+1} = \begin{pmatrix} \cos\mu & \sin\mu \\ -\sin\mu & \cos\mu \end{pmatrix} \begin{pmatrix} Q \\ P \end{pmatrix}_n$$

The origin is a fixed point, whose nature is determined by

$$(10)\quad |\mathrm{Tr}\ M| = |2\cos\mu| \begin{cases} \geq 2 & \text{unstable} \\ < 2 & \text{stable} \end{cases}$$

that is, by the type of eigenvalues (real or complex conjugate).

Besides this simple linear theory we should consider, in a complete view, nonlinear effects coming from imperfections, sextupoles, etc., the fact that the beams may be bunched and subjected to phase stability and synchroton oscillations, radiation in e^+e^- rings, which precludes a description of the system as a conservative one and leads to damping and simultaneous excitation of the oscillations.

In colliding beam machines we have beam-beam interaction, a force of electromagnetic nature acting between the two beams each time they meet. We are going to analyze it in the weak strong model: one of the beams (strong beam) is regarded as fixed; the other (weak beam) consists of test particle, moving under the action of the optics and the force from the strong beam, but without any mutual interaction.

There are two typical ways of colliding, with different effects on betatron motion.

I) 1 ring with 2 bunched beams, circulating in opposite directions, head colliding (e^+e^- rings, CERN pp collider).

(11) $$\Delta x = \Delta y = 0$$

(12) $$\Delta x' = \frac{-2 N_B r_e}{\gamma \beta^2} \left(1 - e^{-\frac{r^2}{2\sigma^2}}\right) \frac{x}{\sigma^2}$$

(13) $$\Delta y' = \frac{-2 N_B r_e}{\gamma \beta^2} \left(1 - e^{-\frac{r^2}{2\sigma^2}}\right) \frac{y}{\sigma^2}$$

N_B is the number of the bunches; β, γ the relativistic parameters; r_e the classical radius of the particle; σ the transverse r.m.s. size of the strong beam

(14) $$r = \sqrt{x^2 + y^2}$$

II) 2 rings with 2 unbunched proton beams, intersecting at fixed angle (ISR, ISABELLE).
Only the vertical degree of freedom is involved in the interaction.

(15) $$\Delta x = \Delta x' = 0$$

(16) $$\Delta y' = (2\pi)^{3/2} \frac{\sigma_y}{\beta_y} \Phi\left(\frac{y}{\sqrt{2}\,\sigma_y}\right) |\Delta \nu_y|$$

(17) $$\Phi(u) = \frac{2}{\sqrt{\pi}} \int_0^u e^{-t^2} dt$$

(18) $$\Delta \nu_y = - \frac{\beta_y r_p I}{\sqrt{2\pi}\, e c \gamma \beta^2 \sigma_y \, tg\left(\frac{\alpha_0}{2}\right)}$$

(tune-shift, strenght parameter of the interaction).
σ_y is the r.m.s. vertical size of the strong beam, β_y the twist parameter (both of them are evaluated at the interaction point), r_p the classical radius of the proton, I the current, e the elementary charge, c the speed og light, α_0 the intersection angle[2].

It is the latter case we want to examine, assimilating it to a 1+1-degree of freedom (revolution+ 1 transverse oscillation) Hamiltonian system.

4. Method of normal forms and Birkhoff series.

Let us suppose we have an area preserving map T of the plane of section into itself, with the origin as an elliptic fixed point. In term of the complex variables

$$(19) \qquad \begin{cases} z = q + ip \\ z^* = q - ip \end{cases}$$

we can decompose the map in a linear part and a nonlinear one

$$(20) \qquad T: z \longrightarrow z' = f(z,z^*) = e^{i\alpha} z + \sum_{m=2}^{\infty} f_m(z,z^*)$$

$f_m(z,z^*)$ is a homogeneous polynomial of m degree.
The area conservation implies that

$$(21) \qquad \frac{\partial(f,f^*)}{\partial(z,z^*)} = 1$$

By a change of variables $(z,z^*) \rightarrow (\xi,\xi^*)$, expressed by the function

$$(22) \qquad z = \varphi(\xi,\xi^*) = \xi + \sum_{m=2}^{\infty} \varphi_m(\xi,\xi^*)$$

$\varphi_m(\xi,\xi^*)$ is a homogeneous polynomial of m degree.

$$(23) \qquad \frac{\partial(\varphi,\varphi^*)}{\partial(\xi,\xi^*)} = 1$$

we can try to write the map as a nonlinear normal form (i.e., a rotation of a radius dependent angle)

$$\bar{T} : \xi \longrightarrow \xi' = e^{i\Omega(\xi,\xi^*)}\xi \tag{24}$$

$$\Omega(\xi,\xi^*) = \alpha + \sum_{m=1}^{\infty} \Omega_{2m}(\xi,\xi^*)^m \tag{25}$$

$|\xi|$ is an integral of motion.
It is possible to derive recurrents relations for the coefficients from the diagram:

$$\begin{array}{ccc} (z,z^*) & \longleftrightarrow & (f(z,z^*),\ f^*(z,z^*)) \\ \updownarrow & & \updownarrow \\ (\xi,\xi^*) & \longleftrightarrow & \left(e^{i\Omega(\xi,\xi^*)}\xi,\ e^{-i\Omega(\xi,\xi^*)}\xi^*\right) \end{array} \tag{26}$$

The series for the transformation (Birkhoff series) are generally asymptotic, that is divergent. The asymptotic behaviour has been analyzed[3] for the quadratic map

$$f(z,z^*) = e^{i\alpha}z - \frac{i}{4}e^{i\alpha}(z+z^*)^2 \tag{27}$$

5. Dragt perturbation theory and Lie series.

Let z be the set of the 2n canonical variables $(q_1,\ldots,q_n,p_1,\ldots,p_n)$. Given a dynamical variable f we can define its Lie operator :f: such that

$$:f:\, g = [f,g] \tag{28}$$

for every dynamical variable g.
$[\ ,\ ]$ is the Poisson bracket.
The Lie operator :f: generates the Lie transformation

$$e^{:f:} = \sum_{m=0}^{\infty} \frac{:f:^m}{m!} = f + :f: + \frac{:f:^2}{2!} + \cdots \tag{29}$$

which is shown to be symplectic. An important property of Lie transformations is

$$e^{:f:}g(z) = g(e^{:f:}z) \tag{30}$$

From this relation it follows that f is an invariant function for the transformation generated by :f:.

Any symplectic map (particularly the ones representing time evolution) can be factorized in product of Lie transformations, ordered according to the increasing degree of the generating polynomials. It is possible to combine exponents using the Campbell-Baker-Hausdorff formula

$$e^{s:f:}\, e^{t:g:} = e^{:h:} \tag{31}$$

$$\begin{aligned} h &= sf + tg + \frac{st}{2}[f,g] + \frac{s^2t}{12}[f,[f,g]] + \frac{st^2}{12}[g,[g,f]] + \ldots \\ &= sf + s{:}f{:}\left[1 - e^{-:f:}\right]^{-1}(tg) + \mathcal{O}(t^2) \end{aligned} \tag{32}$$

Both the methods of Lie series in Dragt's formulation and that of normal forms lead to asymptotic series. The formal analogies of these two different approaches are summarized by

$$\begin{cases} T^n \xi = e^{in\Omega(\xi,\xi^*)}\,\xi \\ T^n z = e^{n:h:} \end{cases} \tag{33}$$

Both the methods provide formal invariants of motion (the series which compute them are generally divergent), respectively $|\xi|$ and h (or its generalization h_r, for the resonant case).

6. Applications to the accelerator map.

We are studying the map of the vertical plane of section describing the motion of a particle in the weak beam under the action of the optics and the strong beam interaction (unbunched intersecting proton beams, weak strong case).

a) Method of normal forms.

The aspect of the global map (oscillation + nonlinear interaction) is (we use the dot for the derivative with respect to the angle θ of revolution)

$$\begin{cases} y_{n+1} = \cos\mu\, y_n + \sin\mu \left(\frac{\dot y}{\nu}\right)_n - \frac{\xi}{\nu}\sin\mu\, \psi(y_n)\, y_n \\ \left(\frac{\dot y}{\nu}\right)_{n+1} = -\sin\mu\, y_n + \cos\mu\left(\frac{\dot y}{\nu}\right)_n - \frac{\xi}{\nu}\cos\mu\, \psi(y_n)\, y_n \end{cases} \tag{34}$$

$$\psi(y) = \frac{\sqrt{2}\,\sigma}{y}\int_0^{y/\sqrt{2}\sigma} e^{-t^2}\,dt = 1 + \mathcal{O}(y) \tag{35}$$

(36) $$\xi = -4\pi\nu/\Delta\nu$$

We remove the linear part of the perturbation, modifying the rotation angle of the linearized map from μ to α

(37) $$\cos\alpha = \cos\mu - \frac{\xi}{2\nu}\sin\mu$$

If $|\cos\alpha| < 1$ the origin is still an elliptic fixed point. We define the complex variables (z,z^*)

(38) $$z = y + i\,\frac{\sin\mu}{\sin\alpha}\left(\frac{\xi}{2\nu}\,y - \frac{\dot y}{\nu}\right)$$

The map is now

(39) $$T : z \rightarrow z' = e^{i\alpha}\left\{z + i\,\xi\,\frac{\sin\mu}{\nu\sin\alpha}\left(\frac{z+z^*}{2}\right)\left[\psi\left(\frac{z+z^*}{2}\right) - 1\right]\right\}$$

where the second term in the parenthesis is the nonlinear part, to be included, step by step, in the normal form [4]. If we choose

(40) $$\sqrt{2}\,\sigma = 1$$

σ disappears from the calculations.

b) Lie series.

(41) $$M_1\begin{cases} q_{n+1} = \cos w\, q_n + \sin w\, p_n \\ p_{n+1} = -\sin w\, q_n + \cos w\, p_n \end{cases}$$ (rotation, linear oscillation)

(42) $$M_2\begin{cases} q_{n+2} = q_{n+1} \\ p_{n+2} = p_{n+1} + f(q_{n+1}) \end{cases}$$ (perturbation)

(43) $$f(q) = \frac{4\pi\,|\Delta\nu|}{\sqrt{3}}\int_0^{q\sqrt{3}} e^{-t^2}\,dt$$

The normalization of the variables (q,p) is given by

(44) $$\begin{cases} q = \dfrac{y}{\beta^{1/2}}\,\dfrac{1}{\sqrt{6}\,\sigma} \\[2ex] p = \dfrac{(\alpha y + \beta y')}{\beta^{1/2}}\,\dfrac{1}{\sqrt{6}\,\sigma} \end{cases}$$

(95% of the beam enclosed in the unitary circle of the plane of section). The perturbation is a kick of impulse depending on the coordinate, just like for all the basically localized interactions.

M_1 can be written as $e^{:f_2:}$

$$(45) \qquad f_2 = -\pi\nu\,(q^2+p^2) = -\frac{w}{2}\,(q^2+p^2)$$

and M_2 as $e^{:F:}$

$$(46) \qquad F(q) = \int_0^q f(u)\,du$$

We try to get a representation

$$(47) \qquad M = M_1 M_2 = e^{:h:}$$

where h is a formal invariant of the motion.
Let us introduce the action-angle variables ($\underline{I}$, $\underline{\varphi}$) for the unperturbed system

$$(48) \qquad \begin{cases} q = (2I)^{1/2} \sin\varphi \\ p = (2I)^{1/2} \cos\varphi \end{cases}$$

At first order in $\Delta\nu$

$$(49) \qquad h = -wI + c_0(I) + 2\sum_{n=1}^{\infty} c_n(I)\,\frac{nw}{\sin(nw)}\,\cos\left[2n\left(\varphi+\frac{w}{2}\right)\right]$$

where

$$(50) \qquad c_n(I) = -\frac{4\pi\,|\Delta\nu|}{3}\left(\frac{3I}{2}\right)^n \sum_{m=0}^{\infty}\left(-\frac{3I}{2}\right)^m \frac{(2m+2n+2)!}{m!\,(m+n+1)!\,(m+2n)!}$$

$$(51) \qquad c_0(I) = 2\pi\,|\Delta\nu|\,I \sum_{m=0}^{\infty}\left(-\frac{3I}{2}\right)^m \frac{(2m)!}{m!\,(2m+1)!}$$

h diverges for all the rational values of $\frac{w}{2\pi} = \frac{k}{2l}$ (k,l integers).

Near the resonances we can construct a function h_r with properties similar to h, but defined even at the resonance itself.

$$(52) \qquad \frac{w}{2\pi} = \frac{k}{2l} + \delta \quad , \quad \delta \ll 1$$

$$(53) \quad h_r = -2\pi\delta I + c_0(I) + 2\sum_{n=1}^{\infty} c_n(I)\,\frac{2n\pi\delta}{\sin(nw)}\cos\left[2n\left(\varphi+\frac{w}{2}\right)\right]$$

The series for h converges fast enough if $w/2\pi$ is sufficiently irrational.

The first terms of the invariants, linear in **I**, are constant for the unperturbed motion. The intensity of the resonant features is determined by the weights if the exploding terms 1/sin (nw), proportional to $nc_n(I)$, and it decreases more than exponentially as n increases, while it decrease going close by the origin. The expression of h reflects two symmetries of the trajectory plots:

1) axial symmetry respect to the line $\varphi = \frac{\pi}{2} - \frac{w}{2}$ (it follows from the invariance for time reversal);
2) central symmetry respect to the origin (it follows from the parity of the powers in the expansion of F(q)).

From the expression of h_r we can approximately calculate the fixed points of M^{21} and their nature (elliptic or hyperbolic)[5].

7. Phase trajectory plots.

It is possible to get a lot of informations about the qualitative behaviour of phase trajectories for the system we are considering following the phase flow, that is, iterating the dynamical map from a set of initial points and for a set of parameters.

Here we are going to show some typical plots concerning:

1) two resonant values of (effective) betatron tune, near ½ and ½;
2) an example of chaotic behaviour, exhibiting the spread of a single trajectory;
3) two sets of parameters for two pp rings, ISABELLE and ISR.

q,p are the same variables as in (44).

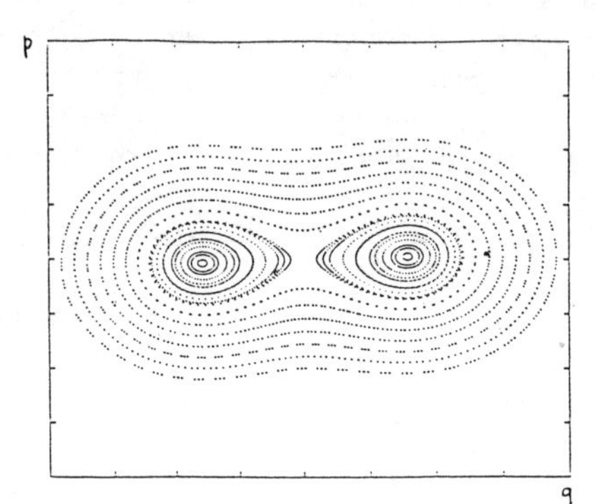

Fig. 4
Trajectory plot for $w/2\pi = 0.5123$ $|\Delta\nu| = 0.01$. q,p range in (−2,2). The origin is a hyperbolic fixed point. There are two islands surrounding two elliptic fixed points.

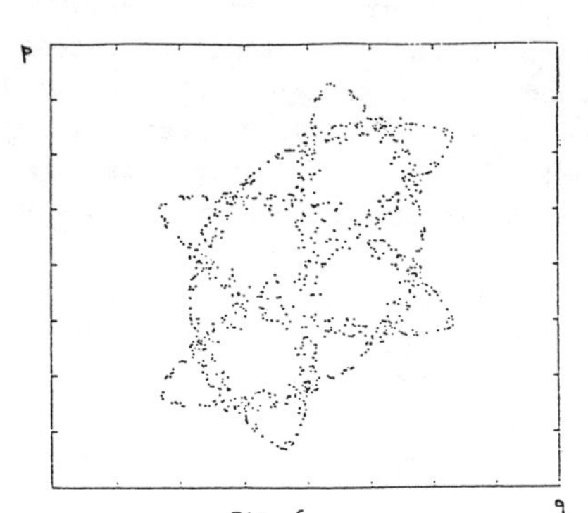

Fig. 6
Chaotic behaviour for $w/2\pi = 0.86$, $|\Delta\nu| = 0.5$. q,p range in (−10,10). The 1000 points belong to the same trajectory, starting from (0.1,0.).

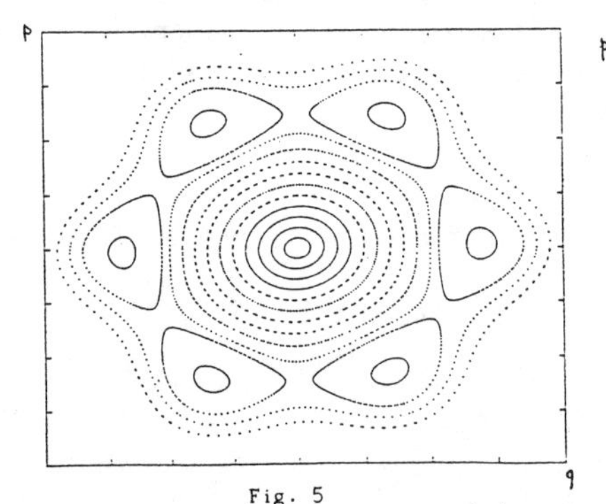

Fig. 5
Trajectory plot for $w/2\pi = 0.1714$, $|\Delta\nu| = 0.01$. q,p range in (−2,2). The origin is an elliptic fixed point. There are six islands surrounding six elliptic fixed points, separated by six hyperbolic ones.

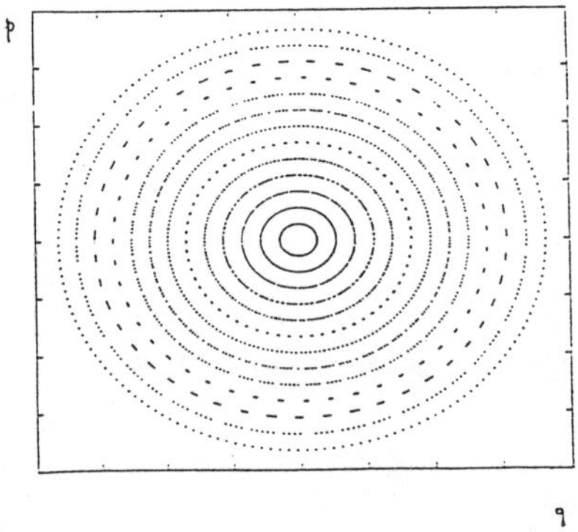

Fig. 7
Regular behaviour for $w/2\pi = 0.7666$, $|\Delta\nu| = 0.0015$ q,p range in (−2,2) (ISABELLE).

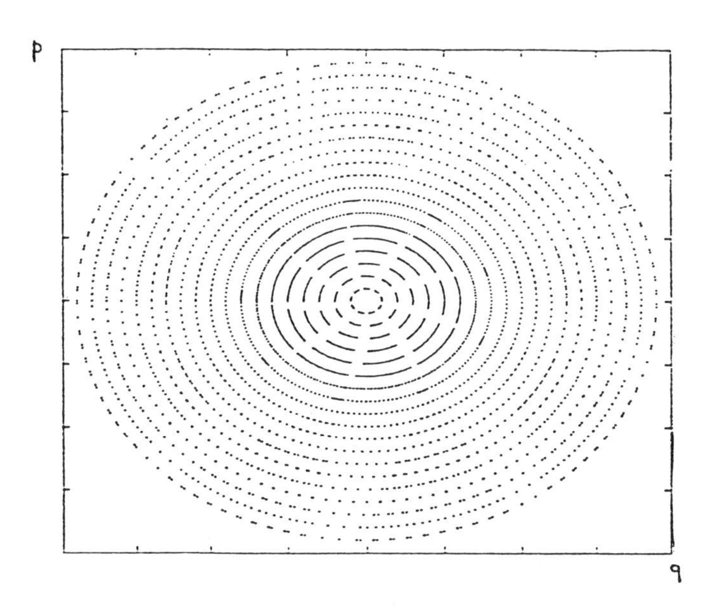

Fig. 8

Regular behaviour for $w/2\pi = 0.1125$, $|\Delta\nu| = 0.0011$.
q,p range in (-2,2) (ISR).

8. Test of the perturbation methods.

We have evaluated the following quantity, estimating the discrepancy between the real dynamics and its simulation by Birkhoff series:

$$\delta = \frac{1}{n}\left\{\left|\,\psi\left(T^n\left(\varphi(\xi_0,\xi_0^*),\varphi^*(\xi_0,\xi_0^*)\right)\right)e^{-in\Omega}\xi_0\,\right|\right\} \tag{54}$$

is the function that generates the inverse transformation $(\xi,\xi^*)\rightarrow(z,z^*)$. The Birkhoff series have been truncated at order 15. We have found a good agreement (δ small enough to allow extrapolations and stability estimates for 10^{11} iterations, that is, for the typical time necessary to store the beam) for ISR. For ISABELLE the agreement is not so good: it is disturbed by the closeness of the resonant value 3/4. The check is very accurate: we control the phase of the complex number, too. The increase of the discrepancy δ as $|\xi_0|$ increases is basically related to the truncation of the series for the error function.

For Dragt method we have evaluated the r.m.s. fluctuation of the function h_0 (invariant for the unperturbed motion) and h and h_r (generalizations using Lie series up to first order in $\Delta\nu$) along an orbit, varying the initial point. We have found that the formal invariants constructed by Lie perturbation series reproduce, with their level lines, the shape of the orbits much better than the unperturbed invariant (emittance or Courant-Snyder invariant). In the resonant case we have found a good agreement all over the range of the initial conditions; in the non-resonant case it is good only in a resticted region, whose radius is about 0.500. There are different causes for the increase of the fluctuation as the radius increases:

- the evaluation is only at first order in the perturbation parameter;
- the series are asymptotic;
- there is an error coming from the truncation of the series for the coefficients $c_0(I)$ and $c_n(I)$ (truncated at order 30).

Acknowledgements.

I am very grateful to M. Pusterla and G. Turchetti for many valuable discussions and suggestions, and to G. Servizi for his precious help in computing problems.

References.

1) R.H.G. Helleman,
Self-Generated Chaotic Behaviour in Nonlinear Mechanics, in Fundamental Problems in Statistical Mechanics, vol. 5, ed. E.G.D. Cohen, North Holland Publ., Amsterdam N.Y. (1980).
2) J.C. Herrera,
Electromagnetic Interactions of Colliding Bemas in Storage Rings, AIP Conference Proc., Vol. 57 (1979).
3) G. Servizi, G.Turchetti, G.Benettin, A. Giorgilli,
Resonances and Asymptotic Behavior of Birkhoff Series, Physics Letters, Vol. 95A, 1 (1983).
4) M. Pusterla, G. Servizi, G. Turchetti,
Nonlinear Beam-Beam Effects in Present and Future Storage Rings. Generalized Birkhoff Method and Conservative Mappings, IEEE Trans. Nucl.Sc., Vol. 30, 4, Santa Fe (1983).
5) A.J. Dragt,
Lectures on Nonlinear Orbit Dynamics, AIP Conference Proc., Vol. 87 (1982).

CHANGE OF ADIABATIC INVARIANT AT SEPARATRIX CROSSING; APPLICATION TO SLOW HAMILTONIAN CHAOS

D.F. Escande

Institute for Theoretical Physics, University of California
Santa Barbara, CA 93106
and
Laboratoire de Physique des Milieux
Ionisés, Ecole Polytechnique,
91128 Palaiseau CEDEX, France

ABSTRACT

For a slowly time-dependent one-degree-of-freedom Hamiltonian system the universal description of the change of adiabatic invariant at separatrix crossing is given and the explicit formula for this change is derived which depends of few parameters corresponding to the specific Hamiltonian of interest. The statistics of this change are derived and previous conjectures on probabilities of capture are proved. In the case of a time-periodic Hamiltonian system the adiabatic invariant is shown to diffuse for orbits which cross the separatrix.

1. Introduction

This paper gives a short and sketchy account of the present state of a work that is being done with J.R. Cary and J.L. Tennyson.[1,2] We consider Hamiltonian systems $\mathcal{H}(p, q, \epsilon t)$ with the following properties:

— They depend slowly on time through parameter ϵ.

— $\mathcal{H}$ frozen at time t has a compact energy surface and a hyperbolic fixed point X.

Therefore the stable and unstable manifolds of X for a frozen $\mathcal{H}$ constitute a separatrix S whose shape is slowly varying with time. Depending on the problem to be considered, S may adopt different shapes shown in Fig. 1.

In the following we focus on the figure 8 shape, but all reasonings can be carried over to the two other shapes by identifying regions a, b, c in the three pictures. As will become obvious later, the rationale for this identification is that in region c an orbit close to S goes twice in the vicinity of X during one period, when it does only once in regions a and b.

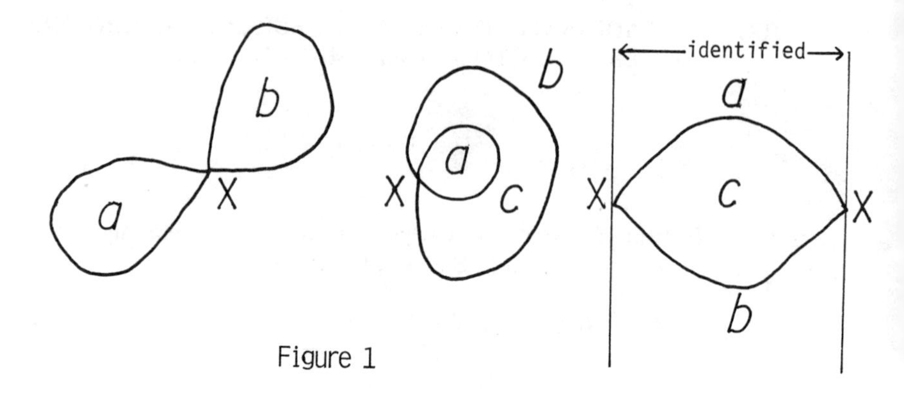

Figure 1

Consider an orbit O of the time-dependent system which has an energy $h(t)$ at time t and is in region α ($\alpha = a,\ b,\ c$) at that time. Let $T_\alpha(h)$ be the period of orbits of the frozen system at time t with energy h in region α, and $\tau = O(1/\epsilon)$ be the characteristic variation-time of $\mathcal{H}$. If

$$T_\alpha[h(t)] \ll \tau, \tag{1}$$

then an adiabatic invariant at order ϵ[3,4] can be defined for O as

$$J_\alpha = I_\alpha[h(t), t] + \epsilon J_1^\alpha[h(t), t, q(t)] + H.O.T \cdot \tag{2}$$

where $I_\alpha(h, t)$ is the action of an orbit with energy h of the system frozen at time t, and $H.O.T$ stands for higher order terms. (I_α is the area trapped inside an orbit of the frozen system, $I_\alpha = \oint pdq$). Notice that condition (1) certainly holds for orbit points far enough from S if ϵ is small enough. It cannot hold close to S for a given ϵ since T_α diverges on S.

Here we are interested in the case where the area enclosed in regions a and/or b is changing with time. For specificity let us consider the case where both lobes a and b are growing with time. Then an orbit initially in region c with adiabatic inviant J_{ic} must cross S at some moment and enter region a or b. Assume it to be

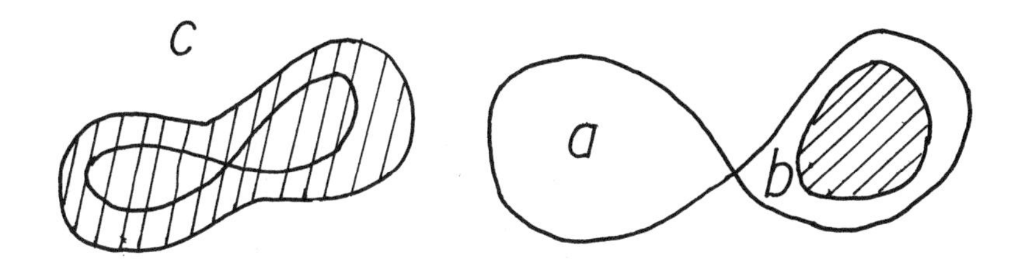

Figure 2

eventually trapped in lobe b (Fig. 2). A new adiabatic invariant J_{fb} can be defined after crossing. We address the following questions.

— If an orbit O is initially in region $\alpha = a,\ b,\ c$ with an invariant $J_{i\alpha}$, and is finally in regions $\beta = a,\ b,\ c$ after crossing S with a new invariant $J_{f\beta}$, what is the value of $J_{f\beta}$ as a function of $J_{i\alpha}$?

— For a set of orbits initially uniformly distributed in the angle conjugate to $J_{i\alpha}$, what are the probabilities of being captured by the two other regions β and γ? What is the probability distribution of $J_{f\beta}$ and $J_{f\gamma}$?

Timofeev gave the answer to the first question for the specific case of a pendulum with a slowly varying amplitude.[5] A conjecture about the answer to the second question was given in refs. 6 and 7. Our theory gives the answer to these questions in the general case. The formula for $J_{f\beta}$ has a universal structure given by Eq. (20) where a maximum of 6 parameters characterizes the specific Hamiltonian under consideration: the exponentiation rate at the fixed point X, an asymmetric parameter, and a maximum of 2 parameters per lobe: the time derivative of the lobe area and a typical energy. If H has enough symmetries the total number of parameters is smaller. It is only two in Timofeev's case[5] and in cases with a similar symmetry [see Eq. (21)]. The probability of capture by a given region is given by Eq. (23), and Eq. (20) together with the equiprobability of h_0 answers the third question.

The answer to these questions finds applications in several fields: transport of particles in plasmas,[2,8] confinement of particles in accelerators, free-electron lasers, celestial mechanics.[7]

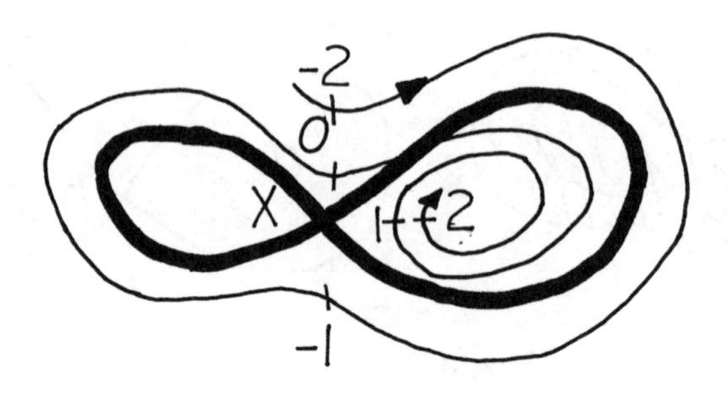

The separatrix is kept artifically fixed, so that the picture looks dissipative

Figure 3

2. Change of adiabatic invariant

2A. O(1) part

For an orbit O which goes from region α to region β, both $J_{i\alpha}$ and $J_{f\beta}$ are $O(1)$ quantities with respsect to ϵ. We now show how to split $J_{f\beta}$ into a trivial to compute $O(1)$ quantity and an $o(\epsilon)$ part which requires careful analysis to be known. Let $I_{s\alpha}(t)$ be the area trapped in lobe $\alpha = a,\ b$ and let $I_{sc}(t) = I_{sa}(t) + I_{sb}(t)$. We define a pseudo-crossing time t_c by

$$I_{s\alpha}(t_c) = J_{i\alpha}. \tag{3}$$

In general t_c is not the time at which O actually crosses S but we may expect the actual crossing time to be close to t_c in some sense. We then define

$$\Delta J_{f\beta} = J_{f\beta} - I_{s\beta}(t_c),$$

which is the previously announced $o(\epsilon)$ quantity. $\Delta J_{f\beta} = 0$ would be the answer if adiabatic theory did not break down close to the separatrix S. Therefore $\Delta J_{f\beta}$ is the measure of non-adiabaticity at separatrix crossing.

2B. Change from vertex to vertex

The slowness of the variation of S implies that O makes many turns close to S before and after crossing it (Fig. 3). We call vertex a point of closest approach to X.

Since no metric is defined this definition depends on the choice of canonical coordinates, but the final result, Eq. (20) does not. We number vertices in the order they are reached by O and give number o to that closest to S in region c. The key point for solving the problem is to compute the change of adiabatic invariant from one vertex to the next one, a method originally used by Timofeev.[5] According to Eq. (2) the order ϵ change of J from vertex n to vertex $n+1$ has a contribution from I and one from J_1. The latter is tedious to compute and can be found in ref. 1. Here only the contribution from I is derived, but this already gives the main ideas of the total derivation.

Let h_n and t_n be the energy and the time corresponding to vertex n of orbit O. The change in action from vertex n to vertex $n+1$ is

$$\Delta I^{\mu}_{n,n+1} = \Big[I_\mu(h_{n+1},t_{n+1}) - I_\mu(h_{n+1},t_n)\Big] + \Big[I_\mu(h_{n+1},t_n) - I_\mu(h_n,t_n)\Big] \tag{4}$$

where μ is the region that contains both vertices n and $n+1$. For n small, t_n is close to t_c and h_n is close to 0 which is taken to be the energy on S. Therefore $[I_\mu(h_{n+1},t_{n+1}) - I_\mu(h_{n+1},t_n)] = \dot{I}_\mu(h_{n+1},t_n)\Delta t_{n,n+1} + H.O.T. = \dot{I}_{s\mu}\Delta t_{n,n+1} + H.O.T.$ and

$$\Delta I^{\mu}_{n,n+1} = \dot{I}_{s\mu}\ \Delta t_{n,n+1} + I_\mu(h_n + \Delta h_{n,n+1}, t_n) - I_\mu(h_n,t_n) + H.O.T., \tag{5}$$

where $\Delta t_{n,n+1} = t_{n+1} - t_n$, $\Delta h_{n,n+1} = h_{n+1} - h_n$, and where $\dot{I}_{s\mu}$, $\mu = a,\ b$ is the time derivative of the area trapped in lobe μ at time t_c and $\dot{I}_{sc} = \dot{I}_{sa} + \dot{I}_{sb}$.

We now turn to the computation of the various quantities in Eq. (5). In order to compute $\Delta t_{n,n+1}$, we can split O between vertices n and $n+1$ into three pieces (Fig. 4): two regions close to X where the trajectory may be approximated by a piece of hyperbola corresponding to the linearized motion, and a third region where O is approximated by a piece of S. This yields

$$\Delta t_{n,n+1} = \frac{1}{\omega}\Big(\ell n\frac{c_1}{|h_n|} + \ell n\frac{c_2}{|h_{n+1}|} + c_3\Big) + H.O.T \tag{6}$$

where ω is the exponentiation rate at X at time t_c, constants c_1 and c_2 are computed through linear theory and c_3/ω is the time necessary to go along S from point A to point B of Fig. 4. Equation (6) holds in particular for a time independent Hamiltonian system. Therefore period $T_\lambda(h)$, $\lambda = a,\ b$ for h small is given by

$$T_\lambda(h) = \frac{1}{\omega}\ell n\frac{e_\lambda}{|h_n|} + H.O.T. \tag{7}$$

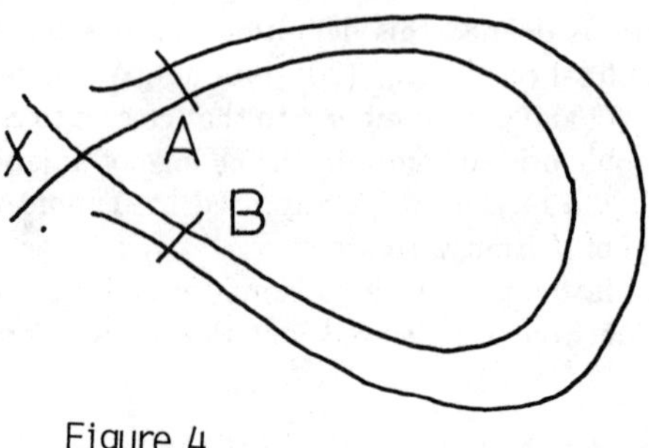

Figure 4

where e_λ is a constant typical of lobe λ. As a result Eq. (6) may be rewritten as

$$\Delta t_{n,n+1} = \frac{1}{2}\left[T_\lambda(h_n) + T_\lambda(h_{n+1})\right] + H.O.T., \tag{8}$$

where λ is the index of the lobe O is moving around or inside form vertex n to vertex $n+1$; ω and e_λ have to be evaluated at time t_c.

The following calculations are made easier by considering (h,t) as conjugate variables, and q as a "time." The corresponding Hamiltonian turns out to be $P(h,t,q)$ as defined as solving $h = \mathcal{H}(p,q,\epsilon t)$ for p. In fact P is a multivalued function of q, but this is no problem since we only use it in integrals along the motion. Hamilton's equations are then

$$\frac{dt}{dq} = \frac{\partial P}{\partial h}, \tag{9}$$

$$\frac{dh}{dq} = -\frac{\partial P}{\partial t}. \tag{10}$$

For the independent Hamiltonian system, Eq. (9) yields

$$T_\lambda(h) = \oint \frac{\partial P}{\partial h} dq = \frac{\partial I_\lambda}{\partial h}, \quad \lambda = a,\ b \tag{11}$$

This allows us to write $I_\lambda(h) = I_{s\lambda} + \delta I_\lambda(h)$ where

$$\delta I_\lambda(h) = \frac{h}{\omega}\left(1 + \ell n \frac{e_\lambda}{|h|}\right) + H.O.T., \ \lambda = a,\ b, \tag{12}$$

and $\delta I_c(h) = \delta I_a(h) + \delta I_b(h)$.

Equation (10) yields

$$\Delta h_{n,n+1} = -\int_{q_n}^{q_{n+1}} \frac{\partial P}{\partial t} dq = -\dot{I}_{s\lambda} + H.O.T., \tag{13}$$

where λ is defined as for Eq. (8); the last term is obtained by approximating the true dynamics by the separatrix one.

Equations (5), (7-8) and (12-13) allow us to compute $\Delta I^\mu_{n,n+1}$ explicitly as a function of h_0 for small $n's$, up to higher order corrections. The result has a logarithmic singularity in h_n. It is correct provided we exclude orbits such that h_{-1}, h_0, or h_1 are of order $\exp -1/\epsilon$.

2C. Expression of the change of invariant

Equation (12) allows us to formally extend the definition of $I_\lambda(h,t)$ for $\lambda = a,\ b$ to small values of h corresponding to region c. In particular we can compute I_λ, $\lambda = a,\ b$ for $h = h_0$. At lowest order J_1^λ, $\lambda = a,\ b$ has a value independent of h at vertices in λ.[1] We can assume it takes the same value at vertex number o. Therefore we can compute J_λ, $\lambda = a,\ b$ at first order in ϵ at this vertex.

Now we can define $\Delta J^\alpha_{-\infty,0} = \sum_{n=-\infty}^{0} \Delta J^\alpha_{n-1,n}$ and $\Delta J^\beta_{0,\infty} = \sum_{n=0}^{\infty} \Delta J^\beta_{n,n+1}$ for any value of the initial lobe parameter α and of the final lobe parameter β taken in $\{a,\ b,\ c\}$. In the sums only the few terms with small $|n|$ contribute appreciably since for higher values of $|n|$, the adiabatic invariance is well perserved. Even though the approximations made for computing $\Delta J^\lambda_{n,n+1}$ are no longer correct for large $|n|$'s, we keep them only for the purpose of getting simple expressions for $\Delta J^\alpha_{-\infty,0}$ and $\Delta J^\beta_{0,\infty}$, and we make an error compatible with our order of approximation. The result is[1]

$$\Delta J^c_{o,\infty} = -(\dot{I}_{sc}/\omega)\big[(2\eta M_c + R_\gamma)\ell n|M_c| - 2\eta M_c - \ell n|\Gamma(1+\eta M_c)$$
$$\Gamma(R_\gamma + \eta M_c)/2\pi|\big], \tag{14}$$

$$\Delta J^\lambda_{0,\infty} = -(\dot{I}_{s\lambda}/\omega)\left[\eta(M_\lambda - \frac{1}{2})\ell n|M_\lambda| - \eta M_\lambda - \ell n\middle|\Gamma(\mu + \eta M_\lambda)\right.$$

$$/(2\pi)^{1/2}\Big]\Big], \ \lambda = a, b, \tag{15}$$

where $R_\lambda = \dot{I}_{s\lambda}/\dot{I}_{sc}$, $M_\lambda = -h_0/\dot{I}_{s\lambda}$, $\eta = +1$, $\mu = 0$, and $\gamma = a, b$ is the lobe O does not visit. $\Delta J^c_{-\infty,0}$ is given by Eq. (14) with $\eta = -1$ and $\Delta J^\lambda_{-\infty,0}$, $\lambda = a, b$ is given by Eq. (15) with $\eta = -1$ and $\mu = 1$.

The adiabatic invariant in region λ at vertex n for $|n|$ small can be written as $J^\lambda_n = I_{s\lambda}(t_n) + \delta J_\lambda(h_n, t_n)$ where

$$\delta J_\lambda(h_n, t_n) = \delta I_\lambda(h_n) + g_\lambda + H.O.T., \ \lambda = a, b \tag{16}$$

$$\delta J_c(h_n, t_n) = \delta J_a(h_n, t_n) + \delta J_b(h_n, t_n) + \dot{I}_{s\nu} \ell n \frac{e_\mu}{|h_n|} - \dot{I}_{s\mu} \ell n \frac{e_\nu}{|h_n|} + H.O.T. \tag{17}$$

where δI_λ is given by Eq. (12), ω and $e_\mu(\mu = a, b)$ are computed at $t = t_c$, $(\dot{\mu}, \nu)$ is defined for O initially in region c (resp. in lobe α) as $\mu =$ final lobe (resp $=$ non -α lobe) and $\nu =$ other lobes (resp. $= \alpha$), and g_λ is the asymmetry parameter of lobe λ, an $O(\epsilon)$ constant given by the ϵJ^λ_1 contribution to the change of adiabatic invariant J_λ [1], which is defined by

$$g_\lambda = -\int_{q_x}^{q_{f\lambda}} dq' \frac{\partial P}{\partial h}(q', o, t_c) \int_{q_x}^{q'} dq" \frac{\partial P}{\partial t}(q", o, t_c) + \int_{q_{f\lambda}}^{q_x} dq' \frac{\partial P}{\partial h}(q', o, t_c) \int_{q_x}^{q'} dq" \frac{\partial P}{\partial t}(q", o, t_c), \tag{18}$$

where q_x is the value of a for the fixed point X at $t = t_c$, $q_{f\lambda}$ is the limit for $h \to 0$ of the q value which equals to $\frac{1}{2}T_\lambda(h)$ the time necessary to go to q from the vertex of an orbit with energy h in lobe λ, and where all integrals are done along the direction of the motion.

Equations (14-15) relate $J_{i\alpha}$ and $J_{f\beta}$ to the value of the adiabatic invariants a time t_0 which depends on the orbit being considered for a given $J_{i\alpha}$. We want to express everything as a function of time t_c. The adiabatic invariant J^α at vertex 0 can be written as $J_{i\alpha} + \Delta J^\alpha_{-\infty,0}$ or as $I_{s\alpha}(t_0) + \delta J_\alpha(h_0)$. Since $J_{i\alpha} = I_{s\alpha}(t_c)$, Taylor-expanding yields

$$t_0 - t_c = \left[\Delta J^\alpha_{-\infty,0} - \delta J_\alpha(h_0)\right] / \dot{I}_{s\alpha} + H.O.T. \tag{19}$$

We can write $J_{F\beta} = I_{s\beta}(t_0) + \delta J_\beta(h_0) + \Delta J^\beta_{0,\infty} + H.O.T.$ and $I_{s\beta}(t_0) = I_{s\beta}(t_c) + \dot{I}_{s\beta}(t_0 - t_c) + H.O.T.$ Therefore $J_{f\beta} = I_{s\beta}(t_c) + \Delta J_{f\beta} + H.O.T.$ where

$$\Delta J_{f\beta} = \Delta J^\beta_{0,\infty} + \Delta J^\alpha_{-\infty,0}\dot{I}_{s\beta}/\dot{I}_{s\alpha} + \delta J_\beta(h_0) - \delta J_\alpha(h_0)\dot{I}_{s\beta}/\dot{I}_{s\alpha}. \tag{20}$$

The first two terms come from the progressive change of the adiabatic invariants when going to and from the neighborhood of the separatrix. The last two terms are the mere consequence of the change in the definition of the adiabatic invariant when going from initial region α to final region β. Parameters g_a and g_b only appear through the global asymmetry parameter $\dot{I}_{sa}g_b = \dot{I}_{sb}g_a$ which is coordinate independent.

The expression of $\Delta J_{f\beta}$ takes a simple form for cases with enough symmetry. Consider the case where $e_a = e_b$, $g_a = g_b$, and $\dot{I}_{sa} = \dot{I}_{sb} = \dot{I}_s > 0$. Then orbits initially in region c are eventually trapped in one of the lobes and we find

$$\Delta J_{f\beta} = -\frac{\dot{I}_s}{\omega}\ell n\left|2\sin\left(\frac{\pi h_0}{\dot{I}_s}\right)\right|, \ \beta = a, b \tag{21}$$

where no e_λ or g_λ appear. For Hamiltonian

$$\mathcal{H}(p,q,t) = \frac{1}{2}p^2 + A(t)\cos\ q, \tag{22}$$

with $\dot{A}(t) < 0$, Eq. (21) may be used[1] with $\dot{I}_s = -4\dot{A}\ A^{-1/2}$ and $\omega = A^{1/2}$. It turns out to be exactly Timofeev's formula.[5] A somewhat less compact formula than Eq. (21) can be given for the case $e_a = e_b$, $g_a = -g_b$, and $\dot{I}_{sa} = -\dot{I}_{sb}$

3. Statistics

$\Delta J_{f\beta}$ as given by Eq. (20) depends on the orbit through $J_{i\alpha}$ (remember that t_c is a function of $J_{i\alpha}$ only) and h_0. Therefore the statistics of $\Delta J_{f\beta}$ is known provided the statistics of h_0 is. This is easily done as follows. Consider at time t_i a set of trajectories uniformly distributed between $J_{i\alpha}$ and $J_{i\alpha} + \delta J_{i\alpha}$ and in the domain $0 \leq \theta_i \leq 2\pi$ where θ_i is the angle canonically conjugate to $J_{i\alpha}/2\pi$ (Fig. 5a).

For simplicity we assume that $\alpha = a$ or b and that $I_{s\gamma}$, $\gamma = b$ or a is not increasing so that all these initial conditions (except possibly for a set of measure $exp - 1/\epsilon$) lead eventually to points in region $\beta = c$. The generalization of the following reasoning is easy to do. The canonical transformation which goes from $(J_{i\alpha}/2\pi, \theta_i)$ to (h_N, t_N) maps the dashed set of Fig. 5a into those of Figs. 5b and 5c which correspond respectively to $N \ll -1$ and to $N = 0$. The limiting curves $t_0(h_0)$ in Fig. 5c are given by Eq. (19) which shows that they are merely deduced

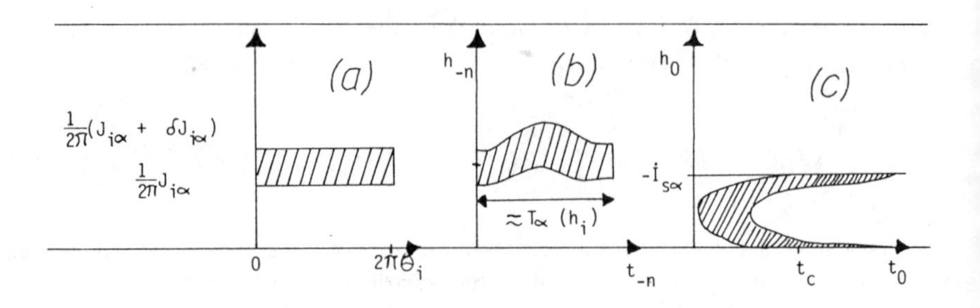

Figure 5

one from the other by a translation $\delta t_c + O(\epsilon \delta t_c)$ where δt_c is the change of t_c, as defined by Eq. (3), when $J_{i\alpha}$ becomes $J_{i\alpha} + \delta J_{i\alpha}$. As a consequence a domain of width Δh_0 between the two limiting curves $t_0(h_0)$ has at lowest order in ϵ and area $\delta t_c \Delta h_0$. Therefore h_0 is equiprobable between 0 and $-\dot{I}_{s\alpha}$ (this bound is given by the definition of vertex 0 and Eq. (13)]. This proves the conjecture of ref. 7. The generalization of this reasoning to the case where the sets of figs. 5a and b can split into two sets in regions β and γ, immediately shows that the probability of capture by each region is proportional to the amout of phase space area gained by each region. If $\dot{I}_{s\beta} > 0$ and $\dot{I}_{s\gamma} > 0$, $\beta \neq \gamma$, orbits initially in region $\alpha (\alpha \neq \beta,\ \alpha \neq \gamma)$ are eventually trapped in regions β and γ with a probability of capture

$$P_\lambda = \frac{\dot{I}_{s\lambda}}{\dot{I}_{s\beta} + \dot{I}_{s\gamma}}, \qquad \lambda = \beta, \gamma. \tag{23}$$

This proves conjectures of refs. 6 and 7.

Since the statistics of h_0 are known, so are those of $\Delta J_{f\beta}$. For the symmetric case $e_a = e_b$, $g_a = g_b$ and $\dot{I}_{sa} = \dot{I}_{sb} = \dot{I}_s > \dot{0}$, it results from eq. (21) that the first two moments of $\Delta J_{f\beta}$ are

$$\langle \Delta J_{f\beta} \rangle = 0, \tag{24}$$

and

$$\langle \Delta J_{f\beta}^2 \rangle = \frac{1}{12} \left(\frac{\pi \dot{I}_s}{\omega} \right)^2 \tag{25}$$

4. Slow Hamiltonian Chaos

We now consider the case where $\mathcal{H}$ is slowly dependent on time but is also periodic. The separatrix makes slow oscillations in a given part of phase space R. The numerical calculation[9] of orbits of Hamiltonian (22) with

$$A(t) = A_0 \cos \epsilon t \tag{26}$$

shows that orbits in R are chaotic. We can show this to be consistnt with KAM theory.

Consider a Hamiltonian $\mathcal{H}(p, q, \epsilon t)$ which is at least of class C^4 (4 times continuously differentiable) in all variables p, q, and t. Rewrite it in the action-angle variables of $\mathcal{H}$ frozen at time t. This yields a new Hamiltonian

$$\mathcal{H}'(I, \theta, \epsilon t) = \mathcal{H}_0(I, \epsilon t) + \epsilon V(I, \theta, \epsilon t), \tag{27}$$

where the second term is due to the time dependence of the canonical transformation. We can embed the (I, θ) motion in an extended phase space (I, θ, w, y) where the motion is governed by Hamiltonian

$$\mathcal{H}"(I, \theta, w, y) = \epsilon w + \mathcal{H}'(I, \theta, y). \tag{28}$$

Obviously $\dot{y} = \partial\mathcal{H}"/\partial w = \epsilon$ and $y = \epsilon t$. We can now go to the action angle variables (J, ϕ) of the (w, y) dynamics governed by Hamiltonian $\epsilon w + \mathcal{H}_0(I, y)$. This yields a new Hamiltonian

$$\mathcal{H}'''(I, J, \theta, \phi) = \mathcal{H}_0(I, J) + \epsilon W(I, J, \theta, \phi), \tag{29}$$

which looks as being of the KAM type. In fact the successive canonical transformations carry over the C^4 differentiability of $\mathcal{H}$ to $\mathcal{H}'''$ only if (p, q) is not in the region R spanned by the slowly varying separatrix. Therefore KAM theorem which requires the C^4 differentiability, tells us that for ϵ small enough there are KAM tori outside of R, but it does not make any statement about the interior of R.

Since $\Delta J_{f\beta}$ depends strongly on h_0 and since, for ϵ small, successive crossings are separated by a large number of turns in phase space, the successive values of $\Delta J_{f\beta}$ are likely to be decorrelated. This leads us to the idea that in the limit of slow Hamiltonian chaos, the adiabatic invariant makes a random walk which can be described by a diffusion coefficient D_{tr} (the index tr stands for trapping) that scales like $\langle \Delta J_{f\beta}^2 \rangle / \Delta t$ where Δt is the time between two separatrix crossings. For the case of the standing-wave Hamiltonian defined by eqs. (22) and (26), by taking into account eq. (25) and $\Delta t = 0(1/\epsilon)$ on obtains[2]

$$D_{\text{tr}} = O(\epsilon^3), \tag{30}$$

and $\partial D_{tr}/\partial A_0 = 0$. This shows that the chaotic transport of J is quite slow. Considered in (p,q) space, there is a part of the chaotic transport which is faster and non diffusive. It corresponds to the fact that orbits with a positive value of p for a time τ such that $\cos\tau = 0$, which are trapped when $t = \tau + \pi/2\epsilon$, are detrapped before $t = \tau + \pi/\epsilon$ with equal probability with $p > 0$ and $p < 0$. Therefore going from $p > 0$ to $p < 0$ takes only a time $O(1/\epsilon)$, when $|p|$ diffuses on a typical time-scale $\epsilon^{-3/2}$.

The diffusion of the adiabatic invariant has to be contrasted with the diffusion of p which occurs slightly after the threshold of break-up ($A_0 \simeq \epsilon^2/8$) of the last KAM surface. In that case the diffusion coefficient for p is given by quasi-linear theory[10] and scales like

$$D_{QL} \propto A_0^2/\epsilon. \tag{31}$$

There is a cross-over between the two regions of chaotic transport for $\epsilon \simeq A_0^{1/2}$

5. Conclusion

We have shown how to compute the change of adiabatic invariant at separatrix crosing for Hamiltonians of the type $\mathcal{H}(p,q,\epsilon t)$. This change is made up of a trivial $O(1)$ part and of a nontrivial part which is $O(\epsilon \ell n \epsilon)$ in general and $O(\epsilon)$ in sufficiently symmetrical cases. The nontrivial part depends on a maximum of six parameters typical of the Hamiltonian being considered. For a given initial invariant, the change strongly depends on a typical energy of the orbit of interest which is uniformly distributed on an interval of order ϵ. If $\mathcal{H}$ is time periodic, the adiabatic invariant is likely to diffuse with a diffusion coefficient which is computable from our theory.

The limit of slow Hamiltonian chaos corresponds to a fairly uniform chaos without visible islands. (at least not too close to the first KAM torus that bounds the chaotic domain). Furthermore far from separatrix crossing, the bent domain of fig. 5c is a straight one (fig. 5b). Therefore chaos seems to be related to the folding of phase space, what is reminiscent of Smale's horseshoe. This gives some hope to prove the decorrelation of successive separatrix crossings for small ϵ.

ACKNOWLEDGEMENTS

Professor A. Kaufman made the useful suggestion to check that our result is coordinate-independent. This research was supported by funds from the National Science Foundation under Grant No. PHY82-17853, supplemented by funds from the National Aeronautics and Space Administration.

REFERENCES

(1) Cary, J.R., Escande D.F., and Tennyson J.L., *Breakdown of Adiabatic Invariance due to Separatrix Crossing*, in preparation.

(2) Cary J.R., Escande D.F., and Tennyson J.L., *Diffusion of Particles in a Slowly*

Modulated Wave, Institute for Fusion Studies report 155, Austin, Texas, 1984.

(3) Lenard A., Ann. Phys. (NY) **6**, 261 (1959).

(4) Kruskal M., J. Math. Phys. **3**, 806 (1962).

(5) Timofeev A.V., Zh. Eksp. Theor. Fiz. **75**, 1303 (1978) [Sov. Phys. JETP **48**, 656 (1978)].

(6) Dobrott D., and Greene J.M., Phys. Fluids **14**, 1525 (1971).

(7) Henrard J., Celes. Mech. **27**, 3 (1982).

(8) Aamodt R.E., Phys. Fluids **15**, 512 (1972).

(9) Menyuk C.R. *Particle Motion in the Field of a Modulated Waves*, to be published.

(10) Lichtenberg A.J. and Lieberman M.A., *Regular and Stochastic Motion* (Springer, New York, 1983) section 5.4.

FREE ENERGY OF SOLITON SYSTEMS IN THE PRESENCE OF SOFT MODES

R. Giachetti*, P. Sodano', E. Sorace°, V. Tognetti"

* Dipartimento di Fisica dell'Università degli Studi di Firenze and Istituto Nazionale di Fisica Nucleare, Largo E. Fermi, 2, I-50125 Firenze, Italy.

' Dipartimento di Fisica e sue Metodologie per le Scienze Applicate dell'Università degli Studi di Salerno, I-84081 Baronissi, Salerno, Italy and Istituto Nazionale di Fisica Nucleare, Mostra d'Oltremare - Pad. 20, I-80125 Napoli, Italy.

° Istituto Nazionale di Fisica Nucleare, Largo E. Fermi, 2, I-50125 Firenze, Italy.

" Dipartimento di Fisica dell'Università degli Studi di Firenze and Gruppo Nazionale di Struttura della Materia del Consiglio Nazionale delle Ricerche, Largo E. Fermi, 2, I-50125 Firenze, Italy.

ABSTRACT

The statistical mechanics of a Sine-Gordon system on a finite support and of a Double-Sine-Gordon are calculated in the semiclassical approximation. The soliton contribution to the free energy is evaluated. In addition to an accurate treatment of translation modes we discuss the relevance of those modes whose frequency vanishes for certain values of the parameters of the systems. Their behaviour is shown to vary continuously between that of an harmonic oscillator and that of a free particle. Correspondingly the soliton interaction decreases and a dilute gas approximation is obtained.

1. - Introduction.

In recent years several papers have been devoted to the relevance of soliton-like excitations in the realm of statistical mechanics of one-dimensional systems. The influence of this kind of elementary excitations on the thermodynamical transport properties of real systems has been verified in different instances and mainly in one-dimensional magnetic chains. As a matter of fact in TMMC and $CsNiF_3$ the experimental behaviour of the magnetic specific heat cannot be explained in terms of linear excitations (spin waves) only [1]. The cross section of neutron scattering, the form of the spectral lines in NMR and Mössbauer experiments also require the introduction of nonlinear elementary excitations [2].

Although the static properties of one-dimensional systems can be found by the transfer matrix method, the distinction of the contribution of soliton and linear modes to the partition function gives a deeper physical insight in the phenomenon, thus allowing a very fruitful way to deal with the dynamical properties.

The principal methods used to evaluate the free energy of a system in terms of linear and nonlinear excitations are the so-called soliton gas phenmenology [3] and the semiclassical approximation in the path-integral approach [4]. Both of them are used not only for integrable systems possessing solitons in strict and technical sense, but also when solitary waves with energy concentrated in a small region are present. These methods consider soliton-like excitations not only as a gas of free particles, but account also for the interaction of solitons with radiation through the modification of the energy levels and the spectral density of states of the linear modes.

The spectrum of the linear modes is obviously dependent on the particular model and can present bound states in addition to the translation mode and the continuum. The translation mode $\omega_0 = 0$ is always present due to the translation invariance of the system and its correct treatment [3] shows the free-particle like behaviour of the soliton. The possible bound states, associated with internal symmetries of the system, have frequencies which depend on the parameters of the model. Some frequency may vanish in correspondence with certain values of the parameters, signalizing instabilities and changes of symmetry of the system. When the frequencies ω_n of these modes are substantially different from zero, so that $\beta\omega_n \gg 1$, they contribute to the internal energy simply as harmonic oscillators. On the contrary, in the limit of $\beta\omega_n \ll 1$, when approaching the instability region, a special care must be devoted to their treatment in order to recover the correct thermodynamical contribution [6]. In this seminar we discuss in detail two systems sharing these features, namely the Sine-Gordon (SG) on a finite support of varying length and the Double-Sine-Gordon (DSG) [7].

2. The Sine-Gordon system as a finite support.

We define a SG system by the Lagrangian

$$\mathcal{L} = \frac{m^3}{\lambda}\int_{-l/2}^{l/2} dx \left[\frac{1}{2}\varphi_t^2 - \frac{1}{2}\varphi_x^2 - V(\varphi)\right] \qquad |2.1|$$

where m is the mass, λ the coupling constant and l=mL, L being the lenght of the real system; φ ,x,t are dimensionless and the potential is $V(\varphi) = 1 - \cos\varphi$.

The classical partition function corresponding to |2.1| has the form

$$|2.2| \qquad Z = A\int \mathcal{D}\left[\tfrac{m}{\sqrt{\lambda}}\varphi\right] \exp\left(-\beta\mathcal{E}[\varphi]\right)$$

where A is an appropriate dimensional constant and

$$|2.3| \qquad \mathcal{E}[\varphi] = \frac{m^3}{\lambda}\int_{-l/2}^{l/2} dx\left[\frac{1}{2}\varphi_x^2 + 1 - \cos\varphi\right]$$

represents the potential energy of the system.

We shall evaluate the functional integral using the saddle-point approximation and assuming that the local minima of |2.3|, determined by the equation

$$|2.4| \qquad \frac{\delta\mathcal{E}[\varphi]}{\delta\varphi(x)} = -\frac{d^2\varphi}{dx^2} + \sin\varphi = 0 \quad ; \; -l/2 < x < l/2$$

are sufficiently separated to give an additive contribution to the partition function. Imposing the periodic boundary conditions $\varphi(l/_2)-\varphi(-l/_2)=2n\pi$, ($n \equiv$ kink number) the solutions of |2.4| are expressed in terms of elliptic functions with a modulus k which for $n \geqslant 1$, is related to the length of the system by l=2nkK(k), so that $l \rightarrow \infty$ for $k \rightarrow 1$. We denote by $\varphi^{(n)}(x)$ the corresponding solution. For n=0, on the contrary, the only stable solution of |2.4| turns out to be the constant solution $\varphi^{(0)}=0$.

According to the saddle-point method we expand $\mathcal{E}[\varphi]$ up to the second order in a neighbourhood of $\varphi^{(n)}(x)$. The diagonalization of the operator of the second functional derivatives amounts to solving the eigenvalue problem

$$|2.5| \qquad \left[-\frac{d^2}{dx^2} + \cos\varphi^{(n)}(x)\right]\eta_J^{(n)}(x) = \left(\omega_J^{(n)}\right)^2\eta_J^{(n)}(x)$$

Consequently the partition function can be cast in the form

$$|2.6| \quad Z=\sum_{n=0}^{\infty} Z_n = A\sum_{n=0}^{\infty} e^{-\beta\mathcal{E}[\varphi^{(n)}]} \prod_J \int dc_J^{(n)}\, e^{-\frac{\beta m^3}{2\lambda}\left[c_J^{(n)}\omega_J^{(n)}\right]^2}$$

When all $\omega_J^{(n)}$ are nonvanishing, all the integrals in |2.6| are easily calculated and finite. This occurs for the constant solution $\varphi^{(0)}$ which gives the usual radiation factor

$$|2.7| \quad Z_0 = A\prod_J \left[\frac{2\pi}{\beta m(1+\frac{2\pi J}{l})}\right]^{1/2}$$

For $n \geqslant 1$ the term $\mathcal{Z}_n = Z_n/Z_0$ represents the normalized contribution of the n-kink sector. Its actual evaluation requires an accurate study of |2.5|, which turns out to be a Lamé equation and can be investigated through the Flaquet-Bloch theory. We find that the eigenvalue are confined in the two bands $0 \leqslant \omega^2 \leqslant (1/k^2)-1$ and $1/k^2 \leqslant \omega^2 < \infty$, the lower one containing exactly n states $\omega_0^{(n)}, \ldots, \omega_{n-1}^{(n)}$. Therefore we shall write

$$|2.8| \quad \mathcal{Z}_n = e^{-\beta\mathcal{E}[\varphi^{(n)}]}\, e^{-H_n} \left(\frac{\beta m}{2\pi}\right)^{n/2} \cdot T_n(\beta)$$

where the second exponential is due to the states in the upper band, while the last factor is the contribution of the lower band.

The evaluation of $H_n = \sum_{J=n}^{\infty} \omega_J^{(n)} - \sum_{J=0}^{\infty} \omega_J^{(0)}$ can be done in the continuum approximation using the density of states

$$|2.9| \quad g(\omega) = \frac{\omega^2 - E(k)/k^2 K(k)}{\left[(\omega^2+1-1/k^2)(\omega^2-1/k^2)\right]^{1/2}}$$

determined from the rotation number of the Lamé equation. The actual value of H_n and its asymptotic behaviour in infinite length are

$$|2.10| \quad \begin{aligned} H^{(n)} &= -\frac{n}{2}\left[K(k)+\ln\sqrt{1-k^2}\right] + n\ln k + \frac{2}{\pi} n K(k)\left(E'(k) - \frac{\pi}{2}k + \sqrt{1-k^2}\ln k\right) \\ &\quad - \frac{n}{\pi} K(k)\left(k^2 K'(k) - \frac{\pi}{2}\right) \xrightarrow[l\to\infty]{} -n\ln 2 + 4l e^{-l/n} \end{aligned}$$

matching the soliton gas phenomenology when n=1.

The lower band needs a much more careful treatment. In the first place it presents a zero frequency mode $\omega_0^{(n)} = 0$ due to the translational invariance of the system. This is the well known translation mode for which the Gaussian integration makes evidently no sense. However, since the eigenfunction $\eta_0^{(n)}(x)$ is proportional to the derivative of $\varphi^{(n)}(x)$, we observe that, by exciting this single mode, we have

$$\varphi(x) = \varphi^{(n)}(x) + c_0^{(n)}\eta_0^{(n)}(x) \simeq \varphi^{(n)}\left[x + \frac{c_0^{(n)}}{2}\left(\frac{2n}{k}E(k)\right)^{-1/2}\right] \tag{2.11}$$

so that the range of the $c_0^{(n)}$ integration has to be limited to the interval $|c_0^{(n)}| \leq \frac{l}{n}[2n\ E(k)]^{1/2}$, giving a factor

$$T_0^{(n)}(\beta) = l\left[\frac{8m^2}{\lambda}\frac{E(k)}{nk}\right]^{1/2} \tag{2.12}$$

For n=1 and $l \to \infty$ we recover the result obtained by Coleman's treatment (5) of the translation mode. However our procedure admits a generalization for finite and a kink number greater than unity. Indeed for n >1 the modes of the lower band, other than the translation, are related to discrete symmetries of the system. These are no more present in infinite lenght, when each kink can be considered independent and the overall picture is that of a dilute soliton gas (3). As a matter of fact the frequency of each mode of the lower band vanishes for $l \to \infty$ so that again Gaussian integrals are senseless. Consider for instance the case n=2. The second mode of the lower band, $\omega_1^{(2)} = \sqrt{1-k^2}/k$ is related to the harmonic force between the two kinks, due to he boundary conditions. When $l \to \infty$ the restoring force becomes smaller and smaller so that a limitation of the range of the $C_1^{(2)}$ integration, similar to that of the translation mode, is in order. As a result we get

$$T_1^{(2)}(\beta) = \sqrt{\frac{2\pi}{m\beta}}\ \frac{1}{\omega_1^{(2)}}\ \Phi\left(l\Lambda\sqrt{\frac{\beta m^3}{2\lambda}}\,\omega_1^{(2)}\right) \tag{2.13}$$

where $\Phi(x)$ is the error function and

$$\Lambda = \left[\frac{1}{k^3}\left(E(k) - (1-k^2)K(k)\right)\right]^{1/2} \tag{2.14}$$

It appears therefore that the thermodinamical contribution of this mode varies continuously from that of an harmonic oscillator (mean energy k_BT, $\ell \ll 1$) to that of a free particle (mean energy $\frac{1}{2}k_BT$, $\ell \to \infty$). The same arguments can be rephrased for each of the mode of the lower band when $n > 1$.

3. – The Double-Sine-Gordon case.

A DSG system is again described by a Lagrangian of the form (2.1) in which the potential depends upon an external parameter R and reads (8)

$$V(\varphi, R) = 1 - \cos\varphi + (3 + \cos\varphi + 4\cos\frac{\varphi}{2})/\cosh R \qquad |3.1|$$

Thus for $0 \leq R \leq \infty$ we get a continuous family of potentials connecting a 4π-SG (R=0) to a 2π-SG ($R=\infty$).

The statistical mechanics of such systems is developed along the lines established in Sect.2. The static soliton like solutions of this model are

$$\varphi_s(x, R) = 4\tan^{-1}\exp(x+R) \pm 4\tan^{-1}\exp(x-R) \qquad |3.2|$$

where each of the two terms is a well known SG kink. the negative sign gives stable 4π kinks while become stable only in the 2π-SG limit, i.e. for $R \to \infty$.

The spectrum of the linear oscillations around a stable solution (3.2), in addition to the zero mode $\omega=0$ and the continuum $1 \leq \omega \leq \infty$ has a bound state whose frequency, for large values of R, is proportional to exp(-R) thus becoming again a pure translation for $R \to \infty$. Therefore the contribution of the bound state to the partition function must be singled out and we shall write the ratio of one-soliton to vacuum contribution in the form

$$\mathcal{Z}_1 = e^{-\beta\mathcal{E}[\varphi_s]} e^{-H} G_0(\beta) G_1(\beta) \qquad |3.3|$$

where again the three factors-besides the Boltzmann one-respectively correspond to the continuum the translation mode and the bound state. Using a procedure analogous to that described in Section 2 we find

$$e^{-H} = 2\sqrt{2}\,\cosh R \cdot \omega_1(R)\,[1 + 2R/\sinh 2R]^{-1/2} \qquad |3.4|$$

and

$$G_0(\beta) = 4\,[\beta m/2\pi]^{1/2}\,\ell\,[1 + 2R/\sinh 2R]^{1/2} \qquad |3.5|$$

For small R the frequency of the bound state, $\omega_1(R)$ is large enough to allow for a Gaussian integration. This yields

$$G_1(\beta) = [\omega_1(R)]^{-1} \qquad |3.6|$$

For large R the bound state gives rise to an approximate collective coordinate. Indeed (9)

$$\varphi_S(x,R) + c_1\eta_1 \simeq \varphi_S(x, R + c_1/N) \qquad |3.7|$$

Imposing the natural cutoff $|C_1| \lesssim RN$ to the C_1-integration we find

$$G_1(\beta) = \Phi\left[R(8\beta m^3/\lambda)^{1/2}\,\omega_1(R)(1 - 2R/\sinh 2R)^{1/2}\right]/\omega_1(R) \qquad |3.8|$$

When $R \to \infty$ the DSG soliton produces a pair of 2π SG solitons which can be considered as independent (10). Imposing therefore the scale relationship $R=\ell/4$, due to the geometry of the system, and introducing a factor 1/2 to account for the appropriate permutations of the two SG-kinks we recover the right expression for the first term of the expansion of the SG partition function in the dilute gas approximation. Let us conclude observing that, for intermediate values of R, the evaluation of the DSG partition function must take into account excluded volume effects which are expected to produce a non-negligible interaction between DSG solitons (11): therefore the possibility of applying the dilute gas approximation must be carefully investigated.

References

(1) F. Borsa, M.G. Pini, A. Rettori and V. Tognetti,
Phys. Rev. B28, 5173 (1983).

(2) M. Steiner, K. Kakurai and J.K. Kiems,
Z. Phys. B53, 117 (1983).
R.C. Thiel, H. de Graaf, L.J. de Jough,
Phys. Rev; Letters 47, 1415 (1981).

(3) A. Krumhansl and J.R. Schrieffer,
Phys. Rev. B11, 3535 (1975).
J.F. Currie, J.A. Krumhansl, A.R. Bishop and S.E. Trullinger B22, 477 (1980).

(4) R. Rajaraman and M. Raj Lakshmi,
Phys. Rev. B25, 1866 (1982).

(5) S. Coleman
in the Whys of Subnuclear Physics, edited by A. Zichichi (Plenum, New York 1973).

(6) R. Giachetti, E. Sorace and V. Tognetti,
Phys. Rev. B30, 3795 (1984).

(7) R. Giachetti, P. Sodano, E. Sorace and V. Tognetti,
Phys. Rev. B30, 4014 (1984).

(8) S. De Lillo and P. Sodano,
Lett. Nuovo Cimento 37, 380 (1983).

(9) D.K. Campbell, M. Peyrard and P. Sodano
(to be published).

(10) About the Behaviour near the instability of DSG see
J.A. Holyst and A. Sukiennicki,
Phys. Rev. B30, 5356 (1984).

(11) K. Sasaki,
Progr. Theor. Phys. 70, 593 (1983).

REMARKS ON THE KdV EQUATION IN LAGRANGIAN COORDINATES

A. R. Osborne
Istituto di Cosmo-Geofisica del CNR, Torino, Italy

A. D. Kirwan
Departement of Marine Science, University of South Florida, St. Petersburg, Florida

A. Provenzale
Dottorato di Ricerca in Fisica delle Universita' di Torino-Cagliari, Torino, Italy and INFN - sezione di Torino

L. Bergamasco
Istituto di Fisica Generale dell'Universita', Torino, Italy

Abstract

The Lagrangian form of the KdV equation (LKdV) is derived and contrasted with the more conventional Eulerian formulation (EKdV). Wave motion described by the LKdV and its associated inverse scattering transform (IST) are discussed. In the case of a single soliton it is shown that the LKdV partitions the particle motion into a kink soliton in the horizontal and a pulse soliton in the vertical. Finally we discuss briefly the Hamiltonian structure of the LKdV equation.

1. Introduction

In recent years spectacular progress has been made in the study of integrable nonlinear wave equations in both one and two spatial dimensions using the inverse scattering (spectral) transform (IST), see [1],[2],[3],[4]. To our knowledge the IST has been exclusively applied to equations in the Eulerian reference frame (E frame) where the evolution of a system is described as a function of time at fixed spatial points (see [5]). It is widely believed that the E frame is most amenable to an observational program.

An alternative way to describe a fluid dynamical system is in the Lagrangian or L frame (see [5]). Because the solution of a dynamical system in the L frame focuses on the motion of material particles of the medium it has some theoretical appeal. Regrettably studies utilizing the L frame tipically range from trivial [5] to interesting but inconclusive [6],[7]. We know of only one study in which the L frame has been effectively used in studying the evolution of a nonlinear integrable system. This is the Lagrangian analysis of the Boussinesq equation provided by Ursell [8]. That study predated the development of the IST.

2. Particle motion for KdV in Lagrangian coordinates

Assuming irrotational, incompressible flow and appropriate boundary conditions Ursell [8] developed a perturbation solution for unidirectional shallow water wave motion in the L frame. The third order problem was found to yield the Boussinesq equation. Using a similar analysis, and further restricting the motion to rightward moving waves, one finds the LKdV equation:

$$\eta_t + c_o \eta_a + \alpha \eta \eta_a + \beta \eta_{aaa} = 0 \tag{1}$$

$$\eta(a,t) = -h \, \partial[\, x(a,t) - a]/\partial a = -h \, \partial X(a,t)/\partial a \tag{2}$$

where c_o, α and β are real constants (for surface water waves $c_o = (gh)^{1/2}$, $\alpha = 3c_o/2h$ and $\beta = c_o h^2/6$, h is the depth, g is the acceleration of gravity). Equation (1) is the wave equation for $\eta(a,t)$ while (2) defines the field in terms of the derivative of horizontal particle position $x(a,t)$ with respect to its position at rest, a, which may be considered as a label for the particle and in the infinite line problem (which we study here) is defined to occur at $t=-\infty$. The vertical y motion evolves according to

$$y(a,t) = b\,[\,1 + \eta(a,t)/h\,] \tag{3}$$

where b is the vertical position of the particle at $t=-\infty$ Thus the position vector at $t=-\infty$ is given by

$$\vec{x}_o = [\, x(a,t=-\infty)\,,\ y(a,b,t=-\infty)\,] = [a,b]\,. \tag{4}$$

Use of (2) in (1) leads to a dynamical equation for X(a,t):

$$X_t + c_o X_a - (\alpha h/2)\, X_a^2 + \beta X_{aaa} = 0 \tag{5}$$

which we refer here to as the potential KdV (pKdV) equation. Inserting (3) in (1) leads to a dynamical equation for y(a,b,t) which is just KdV with different coefficients than (1):

$$Y_t + c_o Y_a + (\alpha h/b)\, Y Y_a + \beta Y_{aaa} = 0 \tag{6}$$

where $Y = y - b$. Given (4) (i.e. the positions of all the particles in the system at rest at $t = -\infty$) and an initial displacement from the equilibrium position (the initial condition) then $X(a,t)$ evolves in time by (5) and $Y(a,b,t)$ evolves independently in time by (6).

Next we consider a simple example. By setting

$$(7) \qquad \eta(a,t) = \eta_o \operatorname{sech}^2 (Ka - \Omega t)$$

which corresponds to a single soliton solution to (1) (where η_o is the soliton amplitude, K, Ω are the spectral wave number and frequency), we integrate (2) to obtain the single soliton solution to (5)

$$(8) \qquad X(a,t) = x(a,t) - a = (\eta_o L/h) [\, 1 - \tanh(Ka - \Omega t) \,]$$

where $L = K^{-1}$ and where we used (4) to set the limits of integration. Equation (8) is the kink soliton solution to (5). More generally the N-soliton solution to (1) can be used in (2) and the N-kink solution to (5) can be found by integrating (2).

The single soliton solution to (6) for the y coordinate motion is

$$(9) \qquad Y(a,b,t) = y(a,b,t) - b = (b\eta_o/h) \operatorname{sech}^2(Ka - \Omega t)$$

Eliminating the common phase $(Ka - \Omega t)$ between (8) and (9) leads to the equation for the particle orbit

$$(10) \qquad Y - Y_1 = -(hb/\eta_o L^2)\, (X - X_1)^2$$

where

$$X_1 = \eta_o L/h \quad , \quad Y_1 = \eta_o /h$$

Thus the trajectory of a single particle originally at rest with coordinates (a,b) is a parabola in the (x,y) plane. Equation (10) was first found by Munk [9] in an early fundamental study on solitary waves. Orthogonal decomposition of this motion yields a pulse soliton in the y direction and a kink soliton in the x direction. The x and y motions may be considered independent provided they have the same spectrum in the sense of the IST. The x spectrum derives from (1) and (2) while the y spectrum derives from (1) and (3). For a single soliton this corresponds to both components having the same discrete (soliton) wave number.

For the general (N solitons plus radiation) solution we may use for the L frame the same nonlinear spectral methods developed for the E frame. The fact that the approximations which lead to the KdV equation in the E frame give an identical equation for the L frame (just the meaning of the variables is changed) allows for immediate use of methods already existing. The knowledge of the nonlinear spectrum of the wave, obtained trough the solution of the direct scattering transform for equation (1), gives a complete knowledge of particle orbits for every wave motion described by the KdV equation. See [10],[11],[12] for a discussion of nonlinear spectral analysis for the KdV equation.

3. Hamiltonian formulation and phase space properties

The Hamiltonian formulation is well suited to the dynamics described by (5) and (6) (see e.g. [13] for the E frame approach). An appropriate Hamiltonian density is given by

$$H = -c_0 X_a \pi + (\alpha h/6) X_a^2 \pi - (\alpha h/3) X_a \pi^2 + b X_{aa} \pi_a , \tag{11}$$

where X is the generalized coordinate and $\pi = -(1/2)\, \partial X/\partial a$ is the generalized momentum. The equation of motion

$$X_t = \frac{\partial H}{\partial \pi} - \frac{\partial}{\partial a}\frac{\partial H}{\partial \pi_a} \tag{12}$$

leads to (5) while

$$\pi_t = -\frac{\partial H}{\partial X} + \frac{\partial}{\partial a}\frac{\partial H}{\partial X_a} - \frac{\partial^2}{\partial a^2}\frac{\partial H}{\partial X_{aa}} \tag{13}$$

leads to (6). Note that the conjugate momentum $\pi(a,t)$ is easily related to the solution of the LKdV equation: $\eta(a,t) = 2h\, \pi(a,t)$.

In the Hamiltonian approach the IST is wieved as a canonical transformation to action - angle variables, owing to the integrability of the KdV motion. In general one may study perturbations to a pure KdV dynamics by adding higher order terms to equation (1). It is not unreasonable to consider following the motion from integrable to non-integrable using simultaneously phase space methods (e.g. Liapunov exponents) and the nonlinear spectral approach (IST). The link between the two is provided by the Hamiltonian structure of the L frame equations of motion. The advantage of using the nonlinear spectral analysis in place of the more conventional Fourier transform is that the former exactly describes the integrable component of the motion, while Fourier analysis is unable to separate the integrable (nonlinear) effects from the non-integrable (nonlinear) behavior.

4. Summary and conclusions

The discussion above shows how the LKdV equation has lead to new physical insight for this type of wave motion. Specifically the Lagrangian components of the particle motion during the passage of a soliton train are a kink train for the propagation direction and a pulse soliton train for the vertical direction. Such a decomposition has yet to be obtained by solely the EKdV formulation. In the case of a single soliton the two components combine to produce a parabola for the path a particle undergoes during the traversal of the soliton. The Hamiltonian formulation of the LKdV approach provides insite on the behavior of the system in phase space.

We acknowledge CNR, INFN and MPI for continuing financial support. This contribution is a preliminary version of an article to be submitted elsewhere.

References

[1] M.J.Ablowitz and H.Segur, Solitons and the Inverse Scattering Transform (SIAM, Philadelphia, 1981)

[2] F.Calogero and A.Degasperis, Spectral Transform and Solitons (North-Holland, Amsterdam, 1982)

[3] R.K.Dodd, J.C.Eilbeck, J.D.Gibbon and H.C.Morris, Solitons and Nonlinear Wave Equations (Academic Press, London, 1982)

[4] V.E.Zacharov,S.V.Manakov,S.P.Novikov and L.P.Pitayevsky, Theory of Solitons. The method of the inverse scattering problem (Nauka, Moskow, in Russian)

[5] H.Lamb, Hydrodynamics (Cambridge University Press, London, reprinted by Dover Publications Inc., 1945)

[6] W.J.Pierson, J.Geoph.Res., 67, 8 (1962)

[7] M.G.Wurtele, J.Meteorology, 17, 661 (1960)

[8] F.Ursell, Proc.Camb.Phil.Soc., 49, 685, (1953)

[9] W.H.Munk, N.Y.Acad.Sci., 51, 376 (1949)

[10] A.R.Osborne, in Statics and Dynamics of Nonlinear Systems, ed. by G.Benedek, H.Bilz and R.Zeyher (Springer Verlag, Berlin, 1983)

[11] A.R.Osborne, A.Provenzale and L.Bergamasco, Nuovo

Cimento, 5C, 612 (1982)
[12] A.R.Osborne, A.Provenzale and L.Bergamasco, Nuovo Cimento, 5C, 633 (1982)
[13] H.Segur, in Topics in Ocean Physics, ed. by A.R.Osborne and P.Malanotte Rizzoli (North Holland, Amsterdam, 1982)

A STRANGE ATTRACTOR IN THE SPIN MODELS

C. Agnes and M. Rasetti
Dipartimento di Fisica del Politecnico, Torino, Italy

ABSTRACT

The non-linear differential equation for the exact correlation function of the Ising (2-d) and XY Heisenberg (1-d) models is mapped into an Hamiltonian dynamical system on $\mathbb{R}^2$. The corresponding diffeomorphism is expected to have a strange attractor, depending on the singular structure of the tangent bundle manifold. The existence of such an attractor is hinted to as well by both the approximate treatments of the models and the features of the algebra induced by the dynamical system Poisson brackets.

A significant insight in both quantum field theory and statistical mechanics can be gained by the analysis of model systems, simple enough that they might be solved exactly, and yet complex enough to share as many global properties as possible with realistic situations in nature.
A major role in this sense has certainly been played by the spin-lattice models (Ising, Heisenberg and vertex models), whose structure has allowed to clarify some of the relevant features of both the mechanism controlling phase transitions and the integrability property[/1/].
The following general scheme may be drawn from them for the whole class of those quantum one-dimensional (or classical two-dimensional) lattice systems whose dynamical variables are somewhat characterized by a symmetry group G associated with a classical (possibly infinite dimensional) Lie algebra.
To begin with, one constructs local charts of canonical coordinates - which are the ones suitable for quantization - by considering all the elements of the center C of the universal enveloping algebra of G, and the orbit $\mathcal{C}$ of G in the group manifold $\mathcal{M}_G$ corresponding to c_r = const. $(=\gamma_r)$, $c_r \in C$.
$\mathcal{C}$ is naturally made into a symplectic manifold if one induces on it a symplectic structure upon identifying the Poisson brackets as follows.
Let $\{ J_n^{(i)};\ i=1,\dots,m\ ;\ n \in \mathcal{N} \}$ be the set of generators of G , whose commutation relations read

$$[J_n^{(i)} , J_k^{(l)}] = \delta_{n,k} \sum_{s=1}^{m} \alpha_s^{(i,l)} J_n^{(s)} \qquad (1)$$

$\mathcal{N}$ is a denumerable set (possibly coincident with $\mathbb{Z}^d$, where d denotes the dimensionality of the lattice Λ ; the lower index of generators refers indeed to the site in Λ); and $\{\alpha_s^{(i,k)}\ ;\ i,k,s = 1,\ldots,m\}$ are the structure constants of G .
Let moreover $\{x_n^{(i)}\ ;\ i=1,\ldots,m\ ;\ n \in \mathcal{N}\}$ be local coordinates of $\mathcal{M}_G$. One sets

$$\{x_n^{(i)}, x_\kappa^{(\ell)}\}_{P.B.} := \delta_{n,\kappa} \sum_{s=1}^{m} \alpha_s^{(i,\ell)} x_n^{(s)} \tag{2}$$

Notice that in terms of such coordinates, if $c_r = \sum \Gamma^{(r)}_{\{n_j i_j\}} J_n^{(i_1)} \ldots J_n^{(i_k)}$ – where the sum runs over suitable sets of indexes – the symplectic manifold $\mathcal{T}$, defined by the system of equations

$$\sum \Gamma^{(r)}_{\{n_j, i_j\}} x_{n_1}^{(i_1)} \ldots x_{n_k}^{(i_k)} = \gamma_r \tag{3}$$

has a direct-product structure $\mathcal{T} = T \times T \times \ldots \times T$ (N_c factors), where N_c is the number of elements of C and T is a manifold.
Resorting now to a functional realization of the irreducible representations of G in which the infinitesimal generators $\{J_n^{(i)}\}$ are realized as differential operators, one has a parametrization of G in the form/2/

$$x_n^{(i)} = X_i\left(p_n^{(1)},\ldots,p_n^{(L)};\ q_n^{(1)},\ldots,q_n^{(L)};\ \{\gamma_r\}\right) \tag{4}$$

$L = \frac{1}{2} \dim \mathcal{T}$, depending on the set of constants which define the selected orbit $\mathcal{T}$.
In the latter parametrization, the $\{p_n^{(i)}, q_k^{(j)}\}$ are conjugate canonical coordinates,

$$\{p_n^{(i)}, q_\kappa^{(j)}\}_{P.B.} = -\delta_{n,\kappa}\,\delta_{i,j} \tag{5}$$

whose quantization is straightforwardly achieved in terms of Heisenberg relations, which map them into the generators of the Weyl algebra.

An integrable quantum lattice system can then be identified in terms of local transfer matrices, by the general Onsager-Baxter-Yang-Zamolodchikov (OBYZ) method[3].

The (local) transfer matrix $M_n(\vartheta)$, $n \in \mathcal{N}$ is an LxL matrix whose elements are formal power series in some running variable ϑ (say $\vartheta \in \mathbb{C}$), with coefficients valued in G.

The requirement for integrability is that one can find a non-singular S-matrix $\mathfrak{S}$ ($\mathfrak{S}$ is an $L^2 x\ L^2$ scalar matrix) satisfying OBYZ's factorization equations.

Indeed the latter amount to stating that if

$$\mathfrak{S}(\vartheta_1 - \vartheta_2)\{M_n(\vartheta_1) \otimes M_n(\vartheta_2)\} = \{M_n(\vartheta_2) \otimes M_n(\vartheta_1)\}\, \mathfrak{S}(\vartheta_1 - \vartheta_2) \qquad (6)$$

$\forall\, \vartheta \in \mathbb{C}$; $n \in \mathcal{N}$, then the infinite family of operators

$$V(\vartheta) = \mathrm{Tr}\left\{ \widehat{\prod_{n \in \mathcal{N}}} M_n(\vartheta) \right\} \qquad (7)$$

commute in G for all $\vartheta \in \mathbb{C}$.

The desired system of (infinite) commuting local conserved quantities can be obtained explicitly by writing the elements of G in $M_n(\vartheta)$, $n \in \mathcal{N}$, in terms of the generators $\{J_k^{(i)}\}$ and expanding $\ln V(\vartheta)$ as a power series in ϑ.

The combinatorial structure connected with the above construction - which is strictly affine to the Dehn's word structure of the group - has far reaching generality: it embodies both the essence of the quantum and classical inverse scattering method and the isospectral deformation properties; moreover it is the generalized counterpart of the Kostant-Symes-Adler lemma which holds in the framework of the coadjoint representation method.

The analytic properties of $M_n(\vartheta)$ as a function of ϑ on a finite Riemann surface control the whole structure of the model.

The 2-dimensional (classical) Ising model and the 1-dimensional XY Heisenberg model-together with the related vertex models - correspond in the above scheme to the lowest dimensional representation of G $[G \equiv SO(3),\ m=3]$.

In terms of 2x2 Pauli matrices $\sigma^{(\alpha)}$, $\alpha = 1,2,3$ (or alternatively x,y,z),

$$J_n^{(i)} := \mathbb{I}_{2^{n-1}} \otimes \sigma^{(i)} \otimes \mathbb{I}_{2^{|\mathcal{N}|-n}} \qquad (8)$$

where

$$[\sigma^{(\alpha)}, \sigma^{(\beta)}] = 2i \sum_{\gamma=1}^{3} \varepsilon_{\gamma}^{\alpha\beta} \sigma^{(\gamma)}$$

$$\{\sigma^{(\alpha)}, \sigma^{(\beta)}\} = 2\delta_{\alpha,\beta} \mathbb{I}_2 \tag{9}$$

In suitable dimensionless units, the XY Hamiltonian has the form

$$\mathcal{H}_{\mathcal{N}} = -\frac{1}{4} \sum_{n\in\mathcal{N}} \left\{ (1+\gamma) J_n^{(x)} J_n^{(x)} + (1-\gamma) J_n^{(y)} J_n^{(y)} + 2h J_n^{(z)} \right\} \tag{10}$$

where γ is an anisotropy parameter and h a magnetic field parameter.
Before briefly reviewing the elegant diagonalization procedure of $\mathcal{H}_{\mathcal{N}}$ based on a Grassmannian realization of the algebra G – which besides giving us the desired ingredients for our further discussion, proves as well the equivalence of the 1-dimensional XY model and the 2-dimensional Ising model in the thermodynamic limit – let us point out a few suggestive facts.
In the scheme above, one can construct coherent states for the algebra of G, and define thereby a semiclassical approximation to the system.
Thus one can associate to a generic Heisenberg system a classical rotator Hamiltonian, whose dynamical variables are obtained by replacing the angular momentum operators by the operator kernel associated with each of them in a coherent state representation/4/.
Single angular momentum coherent states can e.g. be defined as

$$\begin{aligned} |\omega_n\rangle &:= \exp\left\{ i\vartheta_n \left[J_n^{(x)} \sin\varphi_n - J_n^{(y)} \cos\varphi_n \right] \right\} |J\rangle_n \\ &= \left[\cos\left(\tfrac{1}{2}\vartheta_n\right)\right]^{2J} \exp\left\{ \tan\left(\tfrac{1}{2}\vartheta_n\right) e^{i\varphi_n} \left[J_n^{(x)} - i J_n^{(y)} \right] \right\} |J\rangle_n \\ &= \sum_{M=-J}^{J} \binom{2J}{M+J} \left[\cos\left(\tfrac{1}{2}\vartheta_n\right)\right]^{J+M} \left[\sin\left(\tfrac{1}{2}\vartheta_n\right)\right]^{J-M} \times \\ &\qquad \times \exp\left[i(J-M)\varphi_n \right] |M\rangle_n \end{aligned} \tag{11}$$

where

$$|M\rangle_n = \binom{2J}{M+J} \frac{1}{(J-M)!} \left[J_n^{(x)} - i J_n^{(y)} \right]^{J-M} |J\rangle_n \tag{12}$$

$|J\rangle_n$ and $|M\rangle_n$ are both eigenvectors of $J_n^{(z)}$, normalized in such a way that $J_n^{(z)}|J\rangle_n = J|J\rangle_n$, $J_n^{(z)}|M\rangle_n = M|M\rangle_n$; $M,J \in \frac{1}{2}\mathbb{Z}$.
The kernel connected with any operator $\mathfrak{g}_n$ on the n-th spin space is the function $g(\omega_n)$, $\omega_n = (\vartheta_n,\varphi_n)$ such that

$$\mathfrak{g}_n = \frac{2J+1}{4\pi} \int_{\mathcal{S}} d\omega_n \, g(\omega_n) \, |\omega_n\rangle\langle\omega_n| \tag{13}$$

where $\mathcal{S}$ is the unit sphere $S^{(2)}$ in three dimensions, parametrized by polar coordinates ϑ_n,φ_n, so that $d\omega_n = \sin\vartheta_n \, d\vartheta_n \, d\varphi_n$.
Notice that if $\mathfrak{g}_n$ equals the identity operator, then g=1 ; moreover one has the following table of correspondences:

Table I.

$\mathfrak{g}_n$	$g(\omega_n)$
$J_n^{(x)}$	$(J+1)\sin\vartheta_n \cos\varphi_n$
$J_n^{(y)}$	$(J+1)\sin\vartheta_n \sin\varphi_n$
$J_n^{(z)}$	$(J+1)\cos\vartheta_n$
$\left[J_n^{(x)}\right]^2$	$(J+1)\left[(J+3/2)\sin^2\vartheta_n\cos^2\varphi_n - \frac{1}{2}\right]$
$\left[J_n^{(y)}\right]^2$	$(J+1)\left[(J+3/2)\sin^2\vartheta_n\sin^2\varphi_n - \frac{1}{2}\right]$
$\left[J_n^{(z)}\right]^2$	$(J+1)\left[(J+3/2)\cos^2\vartheta_n - \frac{1}{2}\right]$

The Heisenberg Hamiltonian -including quadrupole interactions (which are forbidden quantum-mechanically in the spin-½ case, and correspond just to an additive constant there) - one can thus write is known to have interesting properties. Besides exhibiting topological soliton solutions, it shows in some cases unexpectedly large random fluctuations, characteristic of a chaotic phase.
For example: the XY Hamiltonian $\mathcal{H}_{\mathcal{N}}$ has - for $h \gg \gamma$ - a set of plane-rotator solutions ($\vartheta_n = \frac{1}{2}\pi$, $\forall n \in \mathcal{N}$) whose low lying equilibrium states are determined by the set of equations[5],

$$\sin(\varphi_{n+1} - \varphi_n) - \sin(\varphi_n - \varphi_{n-1}) = \xi\left[\sin\varphi_n - \eta\sin(2\varphi_n)\right] \qquad (14)$$

ξ and η being constants depending on the model parameters h and γ ($J = \frac{1}{2}$). These equations can be recast in the form of a Poincaré mapping

$$\varphi_{n+1} = \varphi_n + \zeta_n$$

$$\sin\zeta_{n+1} = \sin\zeta_n + \xi\left[\sin\varphi_n - \eta\sin(2\varphi_n)\right] \qquad (15)$$

whose phase portrait shows extended chaotic regions.

If out-of-the-plane solutions are considered as well, the resulting equations are more complicated and the mapping is given in terms of three equations.

There is numerical evidence that in configuration space $(\varphi_n, \dot{\varphi}_n, \vartheta_n)$ – where $\dot{\varphi}_n = -(\sin\vartheta_n)^{-1}\,\partial\mathcal{H}_{\mathcal{N}}/\partial\vartheta_n$ – the system has a strange attractor.

The intersection of the orbits with the manifold $\vartheta_n = 0$ induces in the plane $(\varphi_n, \dot{\varphi}_n)$ a diffeomorphism Φ.

It is quite clear from the numerical results that the strange attractor develops as the closure of the unstable manifolds of the saddles of Φ.

Φ shows some similarities with the diffeomorphism connected with the Duffing equation.

A second suggestion comes from the fact that if one approaches the 2-dimensional Ising model in an improved meanfield version – which consists in taking long-range correlations into account by a competing interaction between second-nearest-neighbours – one is once more led to a second order difference equation /6/.

The latter, denoting by M_r the average magnetization per site, by h the applied magnetic field, by p the ratio between the competing exchange interactions and by T the temperature, reads:

$$M_r = \frac{h}{T} + \frac{1}{T}\tanh M_{r-1} - \frac{p}{T}\tanh M_{r-2} \qquad (16)$$

Such an equation is known to have a strange attractor with fractal character, for $p = \tanh^{-1}(1-T^2)$.

The chaotic state is reached through a sequence of bifurcations (i.e. of period doubling modulated phases) characterized by the universal Feigenbaum exponent δ.

All of this hints to the existence of a dynamical system underlying the exact XY Heinsenberg and Ising models, whose structure should on the one hand reflect the integrability of the models themselves - hence show a strictly deterministic behaviour below criticality - and on the other hand be able to switch on chaos for those values of the phase parameters corresponding to disorder in the sense of statistical mechanics.

Such a dynamical system is indeed there, and its structure seems to have the desired properties.

It can be constructed in the following way. One diagonalizes first $\mathcal{H}_{\mathcal{N}}$ by a Jordan-Wigner transformation[7],

$$\begin{aligned} q_m &= J_0^{(z)} \cdots J_{m-1}^{(z)} J_m^{(x)} \\ p_m &= i\, J_0^{(z)} \cdots J_{m-1}^{(z)} J_m^{(y)} \end{aligned} \qquad m = 0,1,\ldots,|\mathcal{N}|-1 \tag{17}$$

[suitable boundary conditions should be imposed, that we do not consider here in detail, in view of the thermodynamic limit ($|\mathcal{N}| \to \infty$) to be eventually taken].

Due to the anticommutation relations of the Pauli operators $\{\sigma^{(\alpha)}\}$ and the definition of the $\{J_n^{(i)}\}$, one has

$$\{p_m, q_n\} = 0 \quad ; \quad \{q_m, q_n\} = 2\delta_{m,n} \quad ; \quad \{p_m, p_n\} = -2\delta_{m,n} \tag{18}$$

Next the q_m's and p_m's are expanded in running Fourier amplitudes

$$\begin{aligned} q_m &= \frac{1}{|\mathcal{N}|} \sum_{\mu=0}^{|\mathcal{N}|-1} \frac{e^{i(m-1)\vartheta_\mu}}{\sqrt{\Omega_\mu}} \left(\psi_\mu + \psi_{-\mu}^\dagger\right) \\ p_m &= \frac{1}{|\mathcal{N}|} \sum_{\mu=0}^{|\mathcal{N}|-1} \sqrt{\Omega_\mu}\, e^{im\vartheta_\mu} \left(\psi_\mu - \psi_{-\mu}^\dagger\right) \end{aligned} \tag{19}$$

where

$$\vartheta_\mu = 2\pi \frac{\mu}{|\mathcal{N}|} \quad ; \quad \mu = 0,1,\ldots,|\mathcal{N}|-1 \pmod{|\mathcal{N}|} \tag{20}$$

The anticommutation relations for q_m, p_m imply that $\psi_\mu, \psi_\mu^\dagger$ are Fermi fields:

$$\{\psi_\mu, \psi_\nu\} = \{\psi_\mu^\dagger, \psi_\nu^\dagger\} = 0 \quad ; \quad \{\psi_\nu, \psi_\mu^\dagger\} = 2\pi\, \delta_{\mu,\nu} \tag{21}$$

On selecting

$$\Omega_\mu = \sqrt{\frac{(1 - z_+ e^{-i\vartheta_\mu})(1 - z_- e^{-i\vartheta_\mu})}{(1 - z_+ e^{i\vartheta_\mu})(1 - z_- e^{i\vartheta_\mu})}} \tag{22}$$

with

$$z_\pm = \frac{1-\gamma}{h \pm \sqrt{h^2+\gamma^2-1}} \tag{23}$$

the Hamiltonian is brought to diagonal form:

$$\begin{aligned} \mathcal{H}_{\mathcal{N}} &= \frac{1}{4} \sum_{m=0}^{|\mathcal{N}|-1} \left\{ (1+\gamma)\, p_m q_{m+1} - (1-\gamma)\, q_m p_{m+1} - 2h\, q_m p_m \right\} \\ &= \frac{1}{|\mathcal{N}|} \sum_{\mu=0}^{|\mathcal{N}|-1} E_\mu \left(\psi_\mu^\dagger \psi_\mu - \frac{1}{2} \right) \end{aligned} \tag{24}$$

where

$$E_\mu = |\varepsilon_\mu| = \sqrt{(\cos\vartheta_\mu - h)^2 + \gamma^2 \sin^2\vartheta_\mu} \tag{25}$$

being

$$\varepsilon_\mu = \cos\vartheta_\mu - h + i\gamma \sin\vartheta_\mu = \varepsilon^*_{-\mu} \tag{26}$$

From this all the properties of the quantum model ground state can be straightforwardly derived: in particular one can check that in the phase plane (γ, h) there are three critical lines corresponding, respectively, to $\gamma = 0$ and $h = \pm 1$. It was a great achievement of Sato and his coworkers /8/ to show how, in the scaling limit

$$h \sim 1 \pm \varepsilon \gamma M \tag{27}$$

with $\gamma > 0$ fixed, ε a vanishingly small positive quantity ($\varepsilon \sim |\mathcal{N}|^{-1} \to 0$) and M a positive real constant; on denoting by $\vartheta = \varepsilon p$ and $E = \varepsilon \gamma \omega(p)$ the scaled variables ϑ_μ and E_μ respectively, one has - close to criticality - the "relativistic" dispersion relation

$$\omega(p) = \sqrt{M^2 + p^2} \tag{28}$$

Moreover if one defines, in the same limit,

$$x = m\varepsilon \quad ; \quad t = n\varepsilon\gamma \quad ; \quad m, n \in \mathbb{Z} \tag{29}$$

one finds (notice that in this limit $\mathcal{N} \sim \mathbb{Z}$)

$$\begin{aligned} e^{in\mathcal{H}_{\mathcal{N}}} \, J_m^{(x)} \, e^{-in\mathcal{H}_{\mathcal{N}}} &\propto \varepsilon^{1/8} \, \varphi_\pm(x,t) \\ e^{in\mathcal{H}_{\mathcal{N}}} \, J_m^{(y)} \, e^{-in\mathcal{H}_{\mathcal{N}}} &\propto \varepsilon^{9/8} \, \frac{\partial}{\partial t} \varphi_\pm(x,t) \end{aligned} \tag{30}$$

where $\varphi_\pm(x,t)$ are the scaled spin operators of the 2-dimensional Ising model for $T \gtrless T_c$ respectively.
The latter is thus essentially reduced to a massive ($M > 0$) Dirac field theory - in one time and one space dimensions.
The model correlation function is therefore nothing but the two-point function of such a field theory, namely (once more in the thermodynamic limit $|\mathcal{N}| \to \infty$, $L \to \infty$, $|\mathcal{N}|/L = \rho_0$ finite and fixed; L denoting the "volume" of the 1-dimensional lattice Λ)

$$\rho(x,x') = \langle \Psi_0 | \varphi^*(x,0) \, \varphi(x',0) | \Psi_0 \rangle \tag{31}$$

where $|\Psi_0\rangle$ is the normalized ground state (in the sense of Dirac), and we have dropped for simplicity of notation the subscripts $\pm$. By customary

translation invariance argument, ρ is indeed a function of $\xi = |x - x'|$.
After setting $\rho(\xi)$ in the form of a Fredholm minor, upon expanding its kernel with the aid of Wick's theorem, and resorting to the property that the monodromy properties of the integral equation are preserved under deformation of x, x', t, t' considered as parameters (in the scaling limit $(t-t') \to 0$), a differential equation can be obtained for it, whose solution is a combination of Painlevé trascendents[9].
Setting

$$\sigma(\xi) =: \xi \frac{d}{d\xi} \ln \rho(\xi) \tag{32}$$

the equation has the form

$$\left(\xi \frac{d^2\sigma}{d\xi^2}\right)^2 = -4\left[\sigma - \xi \frac{d\sigma}{d\xi} - \frac{1}{2}\left(\frac{d\sigma}{d\xi}\right)^2 + \frac{i}{2}(\nu_1+\nu_2+\nu_3)\frac{d\sigma}{d\xi}\right]^2 + 8i \frac{d\sigma}{d\xi} \prod_{k=1}^{3}\left(\nu_k + \frac{i}{2}\frac{d\sigma}{d\xi}\right) \tag{33}$$

where ν_k, k=1,2,3 are parameters depending on both the scale for ρ - fixed by ρ_0 - and the boundary conditions (ξ = 0 and $\xi \to \infty$). We look now at this equation in the following persective. Consider the dynamical system defined by the variables $A = A(\tau)$, $B = B(\tau)$, with "time" $\tau = -2i\xi$; characterized by the Hamiltonian

$$H = H(A,B) = -AB(\nu_1 - B) + \frac{1}{A}(\nu_2 - B)(\nu_3 - B) + B(\nu_2 - B) - (\nu_1 - B)(\nu_3 - B) \tag{34}$$

and Poisson bracket

$$\{B, A\}_{P.B.} = A \tag{35}$$

Its evolution equations

$$\frac{dA}{d\tau} = \{A, H\}_{P.B.} \quad ; \quad \frac{dB}{d\tau} = \{B, H\}_{P.B.} \tag{36}$$

read respectively

$$\Sigma : \begin{cases} \dfrac{dA}{d\tau} = -2B(A-1)^2 + (A-1)(\nu_1 A - \nu_2 - \nu_3) \\ \\ \dfrac{dB}{d\tau} = -AB(\nu_1 - B) - \dfrac{1}{A}(\nu_2 - B)(\nu_3 - B) \end{cases} \tag{37}$$

It is strightforward to check that the above system is equivalent to the equations (we set $\tau = \ln y$, and $A(\ln y) = \mathcal{A}(y)$, $B(\ln y) = \mathcal{B}(y)$)

$$\frac{d^2\mathcal{A}}{dy^2} = \left(\frac{1}{2\mathcal{A}} + \frac{1}{\mathcal{A}-1}\right)\left(\frac{d\mathcal{A}}{dy}\right)^2 - \frac{1}{y}\frac{d\mathcal{A}}{dy} + \frac{(\mathcal{A}-1)^2}{2y^2}\left[\nu_1^2\mathcal{A} - \frac{(\nu_2-\nu_3)^2}{\mathcal{A}}\right]$$

$$\mathcal{B} = -\frac{1}{2(\mathcal{A}-1)^2}\left[y\frac{d\mathcal{A}}{dy} + (\mathcal{A}-1)(\nu_2+\nu_3-\nu_1\mathcal{A})\right] \tag{38}$$

moreover denoting by $a=a(\xi)$ the function $\mathcal{A}(e^{-2i\xi})$,

$$\sigma(\xi) = \frac{\xi^2}{4a(1-a)^2}\left[\left(\frac{da}{d\xi}\right)^2 + 4a^2\right] - \frac{(1+a)^2}{4a} \tag{39}$$

is indeed the desired solution of eq.(33).

As a side remark, it might be worth noticing that the Poisson bracket (35) is particularly suggestive in its structure.

In fact it generates an algebra which is isomorphic to the subalgebra of sl(2) playing an interesting role in the theory of stochastic processes/10/.

Consider a diffusive process over the real line $\mathbb{R}$, controlled by the Fokker-Planck equation ($p=p(x,t)$ denoting now some probability distribution)

$$\frac{\partial p}{\partial t} = -\frac{\partial}{\partial x}\gamma x p + D\frac{\partial^2}{\partial x^2}p \tag{40}$$

and write it

$$p(x,t) = e^{t\mathcal{L}}\, p(x,0) \tag{41}$$

where

$$\mathcal{L} = -\frac{\partial}{\partial x}\gamma x + D\frac{\partial^2}{\partial x^2} \tag{42}$$

The operators

$$Y = -\frac{1}{2}x\frac{\partial}{\partial x} - t\frac{\partial}{\partial t} - \frac{1}{4} \tag{43}$$

and

$$Z = \frac{\partial}{\partial t} \tag{44}$$

together with

$$X = tx\frac{\partial}{\partial x} + t^2\frac{\partial}{\partial t} + \frac{1}{2}t + \frac{1}{4}x^2 \tag{45}$$

generate sl(2), with commutation relations

$$[Y, Z] = Z \quad ; \quad [Y, X] = -X \quad ; \quad [Z, X] = 2Y \tag{46}$$

clearly. $A \sim Z$, $B \sim Y$.

It is interesting that one can write $\mathcal{L}$ as an element of the subalgebra of sl(2) generated by Y and Z only

$$\mathcal{L} = \gamma\left(2Y + \frac{1}{2}\right) + DZ \tag{47}$$

because $\mathcal{L}$ needs to be evaluated only at t=0, where $\frac{\partial}{\partial t} = \frac{\partial^2}{\partial x^2}$.

Thus the diffusive process can be viewed in this way as the orbit under $SL(2,\mathbb{R})$ of the initial configuration, and A,B can be thought of as the same process on the (dual) tangent bundle.

We can finally discuss the results concerning the dynamical system(37).

There is evidence that Σ has a strange attractor.

The unusual feature that characterizes it is the singular nature of the phase manifold, which is what forces the existence of a complex structure containing a non-attracting hyperbolic set as well as an infinite number of

cycles, which are probably responsible for the onset of chaos.
It is such a degeneracy which somewhat rules out the possibility of defining a Bernoulli shift related to the geometric mapping in the customary way: the dynamics is likely not to be of the Smale type (horse shoe) but more complicated (it is possible that one has to enbed the system in a higher-dimensional phase-space, and that on pull back the shift is indeed a braid-like automorphism).
We conjecture also that - at least for certain choices of the parameters - one may have as well Arnold's diffusion .

Acknowledgements

The authors acknowledge interesting and stimulating discussions with M. Jimbo, T. Miwa, R. Livi, S. Ruffo and H. Hamber.
One of them (M.R.) wishes to thank the Director and the Faculty of the Institute for Advanced Study in Princeton, N.J. where part of this work was done, for the warm hospitality.

References

1 L. Onsager, Phys. Rev. 65, 117 (1944).
B. Kaufman, Phys. Rev. 76, 1232 (1949).
E.H. Lieb, T.D. Schultz and D.C. Mattis, Ann. Phys. 16, 407 (1961).
R.J. Baxter, Ann. Phys. 70, 193 (1972).
L.D. Faddeev and L.A. Takhtadzhan, Uspekhi Mat. Nauk 34: 5, 13 (1979).

2 R. Gilmore, "Lie Groups, Lie Algebras and Some of Their Applications", J. Wiley and Sons, New York, 1974.

3 C.N. Yang, Phys. Rev. Lett. 19, 1312 (1967).
A.B. Zamolodchikov, Commun. Math. Phys. 69, 165 (1979).
P.-A. Vuillermont and M.V. Romerio, J. Phys. C6, 2922 (1973).
R.J. Baxter, Ann. Phys. 76, 1, 25, 48 (1973).

4 A.M. Perelomov, Commun. Math. Phys. 26, 222 (1972).
M. Rasetti, Intl. J. Theor. Phys. 13, 425 (1973).
F.T. Arecchi, E. Courtens, R. Gilmore and H. Thomas, Phys. Rev. A6, 2211 (1972).

5 S. Homma and S. Takeno, Progr. Theor. Phys. 72, 679 (1984).
M. Peyrard and S. Aubry, J. Phys. C16, 1593 (1983).

6 C. Buzano, M. Rasetti and M. Vadacchino, Progr. Theor. Phys. 68, 703 (1982).

C.S.O. Yokoi, M.J. de Oliveira and R.S. Salinas, Phys. Rev. Lett. 54, 163 (1985).

7 T.D. Schultz, D.C. Mattis and E.H. Lieb, Rev. Mod. Phys. 36, 856 (1964).

8 M. Jimbo, T. Miwa, Y. Mori and M. Sato, Physica 1D, 80 (1980).

E. Date, M. Kashiwara and T. Miwa, in "Non-Linear Integrable Systems: Classical Theory and Quantum Theory"; M. Jimbo and T. Miwa, eds.; World Scientific Publ. Co. Pte. Ltd., Singapore, 1983.

M. Jimbo and T. Miwa, in "Integrable Systems in Statistical Mechanics", G. D'Ariano, A. Montorsi and M. Rasetti, eds.; World Scientific Publ. Co. Pte. Ltd., Singapore, 1985.

9 P. Painlevé, Acta Math. 25, 1 (1902).

10 M. Suzuki, Physica 117A, 103 (1983).

CONSERVATIVE-LIKE BEHAVIOUR AND ATTRACTORS IN REVERSIBLE DYNAMICAL SYSTEMS

A. Politi, G.L. Oppo

Istituto Nazionale di Ottica, 50125 Firenze - Italy

and

R. Badii

Physik Institut der Universität, 8001 Zürich - Switzerland

ABSTRACT

A reversible system, modelling a laser device, is shown to display simultaneous conservative and dissipative-like features. Global bifurcations ruling structural changes of the behaviour in finite regions of the phase space are also discussed and shown to be related to the birth of a homoclinic cycle. A local analysis on a generic two-dimensional flow is then performed confirming the results obtained for the numerically integrated model.

1. Introduction

The important role played by reversible systems within the hamiltonian dynamics has been clearly evidenced in past years. Many authors have used the reversibility property to develop a clearer and deeper understanding of many physical systems. The first application goes back to G. Birkhoff (1), who was interested in analysing symmetric periodic orbits. More recently, R. de Vogelaere (2) developed an algorithm in order to simplify the numerical study of closed trajectories. Important theorems (KAM), have been afterwards proven by J. Moser (3) in a geometrical context for reversible systems. A recent detailed study of the Standard map (4) is also to be mentioned.

A further relevant contribution is due to R. Devaney (5) who extended the definition of reversible system to include non-hamiltonian flows. The old implicit one was tantamount to saying that the Hamiltonian has to be written as the sum of a quadratic kinetic energy and a potential function. The new one, instead, simply requires the flow to be invariant under the composition of time reversal plus a suitable involution R (i.e. a diffeomorphism such that R^2 = Identity). As it is immediately seen, any system satisfying the former condition is R-reversible with R defined as

$$R(q,p) = (q,-p) \qquad 1.1$$

Here, instead, we provide a model, for a a laser with injected signal with position-dependent Lie derivative and displaying global bifurcations from conservative to dissipative behaviour and vice-versa. As a consequence, we show how the phase space is, in suitable ranges of parameter values, partitioned in distinct regions with structurally different behaviours, separated by stable homoclinic cycles. A few words now to comment on the "conservative" and "dissipative" concepts here used with a slightly wider meaning than usual. The former is not just intended as instantaneously conservative, i.e. an identically zero Lie derivative L_d of the vector field, but it is instead referred to the asymptotic time-averaged value

$$\langle L_d \rangle = \lim_{T \to \infty} \frac{1}{T} \int_0^T L_d(t)\, dt \qquad 1.2$$

which has to be zero. At variance, the term "dissipative" is used as synonimous of the existence of at least an attractor.

In Section 2, we introduce the physical model and discuss its properties when a control parameter is varied.

In Section 3, a generic two-dimensional flow is defined to analyse, in a sympler context, analogous features to those displayed by the laser model in Sec. 2.

Section 4 is finally devoted to conclusions.

2. The physical model and its investigation

The dynamics of a large class of lasers is described by the so called rate equations (6) which relate the energy trapped in the atomic medium (population inversion) with the electric field amplitude.

$$\begin{aligned} \mathring{E} &= wE \\ \mathring{w} &= d - E - \varepsilon(1 + E^2)w \end{aligned} \qquad 2.1$$

here written in the adimensional time scale $\tau = (k\,\gamma_{\parallel})^{1/2}$, with $k, \gamma_{\parallel}$ being the relaxation rates for the field E and the population inversion w, respectively. Let us also recall that w has been, for simplicity, referred to its equilibrium value, while ε is defined as follows

$$\varepsilon = \sqrt{\gamma_{\parallel}/k} \qquad 2.2$$

and, finally, d is the pump parameter referred to the threshold value. The local volume contraction-rate L_d, is given, in the space (E,w), by

$$L_d = w - \varepsilon(1 + E^2) \qquad 2.3$$

and we see that, while the meaning of the second term in the r.h.s. clearly stands out as a dissipation, the first term is either positive or negative, depending on the sign of w. However, a more careful choice of variables allows a clear-cut discrimination. Namely, the introduction of

$s = \ln E$, leads to the following equation

$$\mathring{\mathring{s}} = -\varepsilon\mathring{s}\,(1 + \exp(2s)) + d - \exp(2s) \qquad 2.4$$

definitely proving that the laser equations are equivalent to a damped anharmonic (Toda) oscillator (7), with the ε-term representing the only truly dissipative one. As a consequence, those lasers like CO_2 or Nd-Yag, where $\varepsilon \ll 1$, are approximately described by a hamiltonian flow.

Let us now consider such a class of lasers when they are externally driven by another, sligthly detuned, laser field. In this case, a third equation referring to the relative phase ϕ between the two laser signals must be added and the resulting model is

$$\begin{aligned} \mathring{w} &= d - E^2 - \varepsilon(1 + E^2)w \\ \mathring{E} &= wE + E_o \cos\phi \qquad 2.5 \\ \mathring{\phi} &= -\theta - (E_o/E)\sin\phi \end{aligned}$$

where θ and E_o are the cavity mistuning and the Rabi frequency of the external field respectively, both normalized to the present time scale. It is immediately seen that the asymptotic Lie derivative remains substantially unchanged

$$\langle L_d \rangle = \lim_{T\to\infty} \frac{1}{T}\left\{\int_0^T (2w - \varepsilon(1 + E^2))dt - \int_0^T (\mathring{\ln} E)dt\right\} \qquad 2.6$$

since the last term gives a null contribution, being $\ln E(t)$ bounded both from below and above. Moreover, in the limit case $\varepsilon = 0$ the model 2.5 is reversible. Let us first rewrite the equations

$$\begin{aligned} \mathring{r} &= zr + f\cos\phi \\ \mathring{z} &= D - r^2 \qquad 2.7 \\ \mathring{\phi} &= -1 - (f/r)\sin\phi \end{aligned}$$

where time has been rescaled by θ, $r = E/\theta$, $z = w/\theta$ and the new parameters are $f = E/\theta^2$ and $D = d/\theta^2$. Now, the flow 2.7 is straightforwardly seen to be invariant under time reversal plus the involution R

$$R(x,y,z) = (-x,y,-z) \qquad 2.8$$

written in cartesian rather than cylindric coordinates. Such a property is not, however, sufficient to guarantee the existence of a suitable nonlinear change of coordinates leading to a locally null L_d, as for the unforced laser case. In fact, an additional hypothesis must be generically verified by a trajectory, in order to be conservative: symmetry with respect to R-involution. In this case, the same dynamical properties displayed forward in time by the trajectory $\{\underline{x}_t\}$ would be, indeed, shared by backward evolution of its symmetric $R\{\underline{x}_t\}$ and hence, they would have, for instance, reversed Lyapunov spectra. Therefore, if $\{\underline{x}_t\}$ is R-invariant, its spectrum turns out to be antisymmetric with respect to the 0 value as for a hamiltonian flow. When instead, the trajectory is not symmetric, nothing can be, a priori, said on its character.

Coming back to the flow 2.7, we see that, when $f > f_c = \sqrt{D}$, a pair of asymmetric nodes exists, and one of the two (indicated with Q_1)

$$Q_1 :(r_1 = \sqrt{D}, \ \phi_1 = \mathrm{arctg}(1/z_1), \quad z_1 = \sqrt{f^2/D - 1} \ ) \qquad 2.9$$

is attracting.

An important question now arises on the way symmetry-breaking occurs. Indeed, varying f from zero to f_c , the system changes from a pair of uncoupled oscillators with the whole phase space filled up with conservative solutions, to an at least locally dissipative flow. We have thus performed a detailed numerical analysis choosing D = 4/3, fixing an initial grid of 400 points and, then, following the evolution through the successive iterates on a suitable surface of section (discarding the first 100 points in order to possibly estinguish transients and evidence the onset of dissipative structures). Some of our results are reported in Figs. 1-3, where the Poincarè section has been obtained taking the maxima of the field intensity $(x^2 + y^2)$. A first qualitative analysis seems to suggest that a globally conservative structure still persists at f=0.4, while, for f=0.45 a closed region filled with tori and, probably, chaotic layers, coexists with the external basin of attraction of a period-2 cycle. Further studies based on the evaluation of the Floquet exponents lead to the same conclusions on the conservativeness and hence on the symmetry of solutions.

The nature of the bifurcation ruling the onset of this attracting orbit has been then investigated. A first local study reveals an anomalous pitchfork scheme (see Fig. 4). While the middle branch changes its stability property from ellyptic to hyperbolic, the two external branches have opposite stability. Such a scheme is quite degenerate if considered in the broad class of dynamical systems, while we conjecture it is generic within the reversible ones. This point will be clarified in the next section.

The main aspect is, however, the global character of this bifurcation which is related to the birth of a homoclinic cycle. In fact, the trajectory separating the two regions is nothing but the intersection of the stable and unstable manifolds of the hyperbolic period-2 cycle and it is interesting to notice that it persists over a finite region of f-values, that is, it cannot be perturbed away, as it usually occurs.

We end this section by recalling a few simple properties of the two-dimensional map associated to the Poincarè section herein shown.

R-symmetry of the flow is mapped into S-symmetry of the diffeomorphism

$$S(x,y) = (-x,y) \qquad 2.10$$

since the change of sign of z is insured by the condition for a point to belong to the surface of section $(z=-Ax/(x^2+y^2))$. Then, the S-reversibility can be written as

$$S \circ M \circ S = M^{-1} \qquad 2.11$$

where M indicates our map, or, equivalently, $(S \circ M) \circ (S \circ M)$=Identity. Therefore, calling $G = S \circ M$ we have

$$M = S \circ G \qquad 2.12$$

that is, M is the product of two involutions as it happens for the Standard map (4) or the Hénon conservative one (8). However, only a model having a non-constant Jacobian, as the one we have presented, can show birth and death of conservative and dissipative structures.

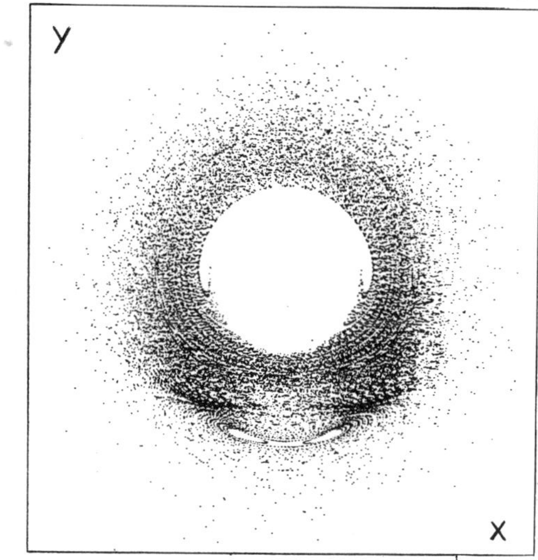

Fig. 1) Poincarè section (r maxima) for D=4/3, f=0.3975. The system appears to be conservative, but no more integrable as for f=0. The central empty region corresponds to minima of r. Both x and y values range between -4 and 4.

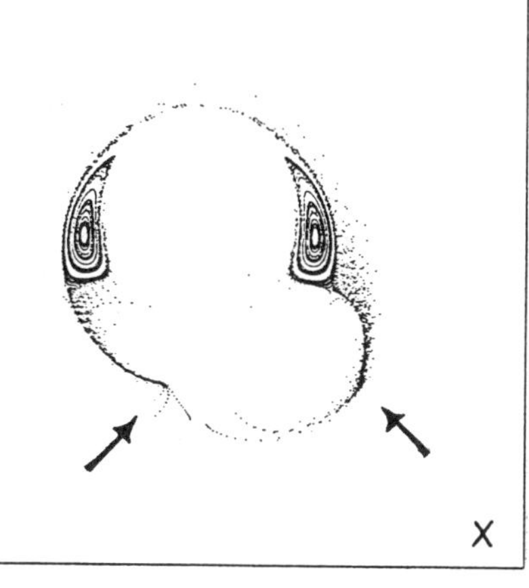

Fig. 2) Same as in Fig. 1 but for f=0.45. The lower region of tori has disappeared, giving rise to a dissipative structure. All initial conditions outside the upper conservative region asymptotically fall onto the period-2 solution shown by the arrows.

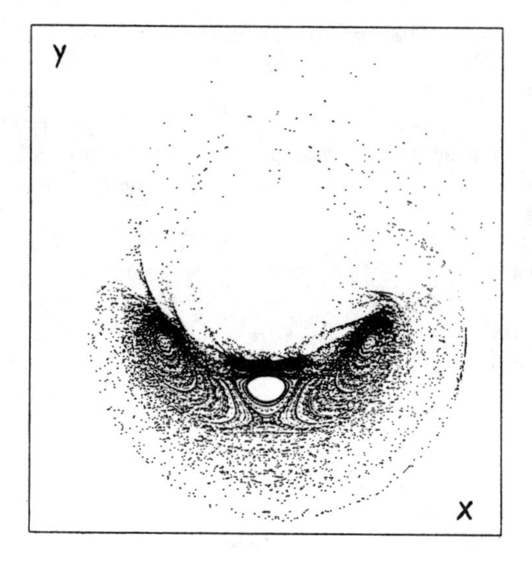

Fig. 3) Same as in Fig. 1, for f=1. Notice the lower conservative region on whose top an asymmetric period-14 solution (hidden by its transient) exists.

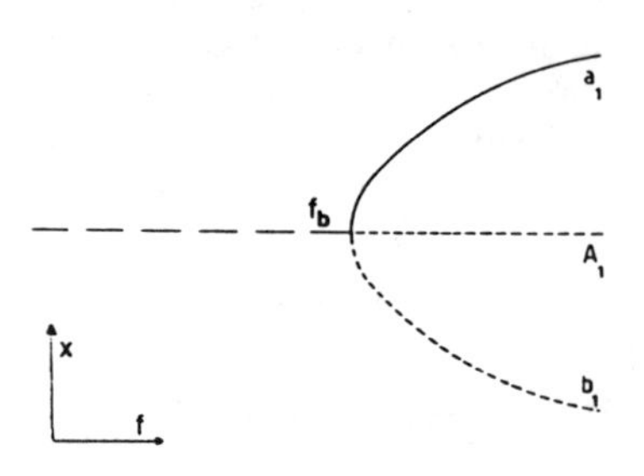

Fig. 4) The bifurcation diagram ruling the onset of the attractor-repellor pair. The middle branch changes from elliptic to hyperbolic.

3. A generic two-dimensional reversible flow

In the previous Section we have shown a laser model displaying the simultaneous presence of conservative and attracting features under realistic assumptions on parameter values.

Here, instead, reducing the number of degrees of freedom from three to two we introduce a general model in order to recover similar results in a more intelligible context. In fact we have to deal with periodic solutions and fixed points instead of tori, lockings and even chaotic layers. The model is built along the standard procedures followed in bifurcation theory. We start with a reversible system corresponding to the linearization of a flow around a symmetric fixed point taken for simplicity in the origin. The symmetry axis of the involution is chosen, without loss of generality, to be y = 0. Hence, a generic flow is written as

$$\mathring{x} = a_1 y \qquad\qquad 3.1$$
$$\mathring{y} = b_1 x$$

and it is immediately seen that, depending on the sign of $a_1 b_1$, the origin is a hyperbolic ($a_1 b_1 > 0$) or an ellyptic ($a_1 b_1 < 0$) fixed point.

Now we add all second order terms which leave the same symmetry axis invariant

$$\mathring{x} = a_1 y + a_2 x^2 + a_3 y^2 \qquad\qquad 3.2$$
$$\mathring{y} = b_1 x + b_2 xy$$

Let us now forget about the local character of model 3.2 and study its global properties. We immediately see that a second symmetric fixed point exists for any choice of parameter values, namely

$$x_1 = 0 \,, \qquad y_1 = -a_1/a_2 \qquad\qquad 3.3$$

which, again, can be either ellyptic if

$$(a_1/a_2)(a_1 b_2 - a_2 b_1) < 0 \qquad\qquad 3.4$$

or hyperbolic in the opposite case, as it can be easily ascertained by evaluating the eigenvalues of the linearized equations

$$\delta\mathring{x} = 2a_2 x_\circ \delta x + (a_1 + 2a_3 y_\circ)\delta y \qquad\qquad 3.5$$
$$\delta\mathring{y} = (b_1 + b_2 y_\circ)\,\delta x + b_2 x_\circ \delta y$$

where x_o, y_o is a generic singular point. Two more asymmetric fixed points can also exist

$$x_{3,4} = \pm\sqrt{(b_1/b_2 a_3)(a_1 b_2 - a_2 b_1)} \,, \qquad y_{3,4} = -b_1/b_2 \,, \qquad\qquad 3.6$$

Thus we recover, with the simplest general assumptions on the non-linearities, a phase-space structure similar to that found in the previous Section. A further analysis of the parameter space shows that if

$$a_1,\ a_3,\ b_1,\ b_2 < 0 \,, \qquad a_2 > 0 \qquad\qquad 3.7$$

the fixed point (x_1, y_1) is ellyptic with a positive ordinate y_1, the origin is hyperbolic and, finally, two asymmetric points with $y^1_{3,4} < 0$ exist, the one with positive abscissa being a stable node.

Under the assumptions 3.7, the phase space is partitioned as qualitatively described in Fig. 5: the region enclosed by the homoclinic orbit consists of periodic solutions in agreement with a theorem proven by R. Devaney.

For what concerns the asymptotic stability at large x's and y's, let us first rewrite eqs. 3.2, neglecting the linear terms, in polar coordinates

$$\begin{aligned} \mathring{r} &= r^3 \cos\phi \, (a_3 \cos^2\phi + (a_2 + b_2) \sin^2\phi) \\ \mathring{\phi} &= r \sin\phi \, ((b_2 - a_3) \cos^2\phi - a_2 \sin^2\phi) \end{aligned} \qquad 3.8$$

Since the loci where $\mathring{\phi} = 0$ (as well as those where $\mathring{r} = 0$) are straight lines defined by ϕ = const., it immediately follows that they represent asymptotically invariant trajectories. Therefore, they have to be connected with the stable and unstable manifolds emananting from the asymmetric fixed points. Under general conditions, $\mathring{r}$ changes sign at each crossing of any of the three straight lines defined by

$$\cos\phi = 0 \quad , \; \mathrm{tg}\,\phi = \pm\sqrt{-a_3/(a_2 + b_2)} \qquad 3.9$$

and, similarly, for $\mathring{\phi}$ on the follwing three

$$\sin\phi = 0 \quad , \; \mathrm{tg}\,\phi = \pm\sqrt{(b_2 - a_3)/a_2} \qquad 3.10$$

We can also see that under the assumptions 3.7 the critical value where $\mathring{r}$ changes sign is always larger than the corresponding one for $\mathring{\phi}$. As a consequence, a global picture as that in Fig. 6 is to be expected, and we see that, except for the triangular sector on the left, the whole phase space is asymptotically stable. It is important to study now how the structure depicted in Fig. 5 changes when some parameter is varied. In particular, we are interested in describing the mechanism through which the stable attractor disappears, yielding a wholly conservative picture. The relevant parameter to control is b_1, whose zero-crossing leads both to the disappearance of the asymmetric solutions and to the change of stability of the origin, which becomes an ellyptic point. In other words we have, as in the previous Section, an anomalous flip-bifurcation which, owing to the strong symmetry conditions for the vector field, turns out to be a generic phenomenon in reversible systems.

To conclude the discussion on the phase-space structure we give a qualitative picture for $b_1 > 0$. We see that two ellyptic regions are coexisting, separated by an unstable region where a flow from $+\infty$ to $-\infty$ is settled. Thus we find again an analogous behaviour to that displayed by the physical model for $f < f_c$ where a point mapped to infinity exists. This situation is displayed in Fig. 7.

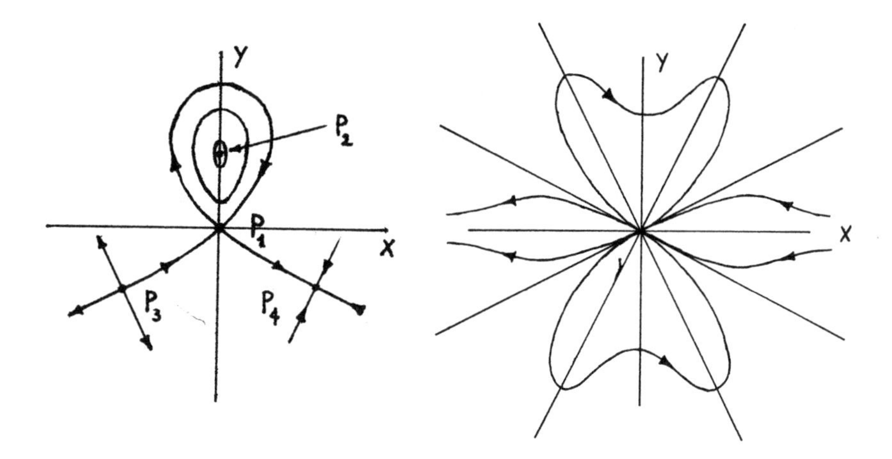

Fig. 5) Qualitative flow diagram for eqs. 3.2, corresponding to the conditions 3.7. Conservative limit cycles, encircling P_2, are enclosed by a homoclinic cycle.

Fig. 6) Large-distance approximation of flow 3.2, for conditions 3.7.

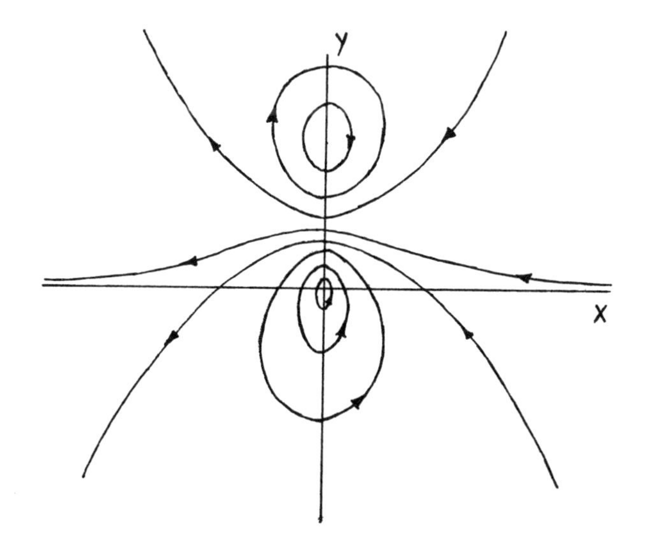

Fig. 7) Same as in Fig. 5, but for b_1 0. Two regions of invariant curves are separated by an unbounded flow.

4. Conclusions

A particular laser system has been analysed as a prototype for a generic reversible system with position-dependent divergence. When a suitable external parameter is varied, global structural changes are observed. Finite regions in the phase space loose their conservative character as a consequence of the appearance of a periodic attracting orbit. The mechanism of this transition has been interpreted in terms of symmetry-breaking of a periodic solution giving rise to a homoclinic cycle.

The distinction between conservative and dissipative behaviour is not purely academic in the CO_2-laser with injected signal. It is, instead, crucial to understand the onset of very stable orbits in the complete model, which includes explicit damping terms.

REFERENCES

1) G. Birkhoff, Rend. Circ. Mat. Palermo 39, 265 (1915).
2) R. de Vogelaere, In Contributions to the Theory of Nonlinear oscillations, Ed. by S. Lefschetz (Princeton University Press, Princeton 1958).
3) J. Moser, Stable and Random Motions (Princeton University Press, Princeton 1973).
4) J.M. Greene, J. Math. Phys. 20, 1183 (1979).
5) R.L. Devaney, Trans. Am. Math. Soc. 218, 89 (1976).
6) M. Sargent III, M.O. Scully and W. E. Lamb jr., Laser Physics (Addison-Wesley, Reading 1974).
7) G.L. Oppo and A. Politi, Z. Phys. B59, 111 (1985).
8) M. Henon, Comm. Math. Phys. 50, 69 (1976).

HYSTERESIS AND ATTRACTOR CRISIS IN A FORCED OSCILLATOR

Giovanni Riela
Istituto di Fisica dell'Università,
Via Archirafi 36 - 90123 Palermo (Italy)

ABSTRACT

A detailed study of a Van der Pol-like oscillator is provided. Special emphasis is given to the onset of chaos via the break-up of an invariant torus. Bifurcation diagrams of periodic solutions are also carefully described.

We have considered a model given by the following autonomous system of three non linear differential equations

$$\dot{X} = .7\ Y + X - 10\ XY^2$$

$$\dot{Y} = -X + R \sin Z$$

$$\dot{Z} = \pi / 2$$

This model, mentioned in Ref. 1, describes a modified Van der Pol oscillator with a driving term sin $(\pi/2.t)$ whose amplitude R is taken as control parameter.

The system is invariant under the symmetry operation associating the point (X,Y,Z) with $(-X,-Y,Z+\pi)$. As a consequence, attractors are ,either symmetric (i.e., invariant under the above defined transformation), or appear in symmetrically conjugate pairs.

We have numerically studied the above system in the parameter range $0 \leqslant R \leqslant .6$. The resulting phenomenology shows an interesting interplay between three groups of periodic orbits, two invariant tori and several chaotic attractors. A global picrture is given in Fig. 1, where the symbol P_j^K indicates the k-th orbit of period jT, and the prime means hyperbolicity For instance, $P_2^{I'}$ is a hyperbolic orbit born, together with P_2^I , through a tangent bifurcation, and symmetrically conjugated to a second orbit P_2^2 of period 2T. Similarly $(CH_j)^K$ indicates the chaotic attractor originated from P_j^K through a Feigenbaum cascade of period doubling bifurcations. T_1 and T_2 indicate two invariant tori which characterize the quasi-periodic dynamics found in the initial and final region of parameter range herein considered.

Hysteresis

In the intermediate region different attractors coexist, each with its own basin of attraction. Looking at Fig. 1 we see that between R_1 and R_3 the P_2 's or their subharmonics, generated through period doubling bifurcations, coexist. Between R_3 and R_4 , we find the strange attractors CH_2^1 produced by the Feigenbaum cascades of the P_2's. Between R_5 and R_6 the chaotic attractor denoted with CH_2, coexists with the symmetric periodic orbit P_3 born at R_5, together with its hyperbolic twin P_3 '. A symmetry -breaking bifurcation of P_{3a} leads, inside the range R_6-R_7, to a pair of distinct solutions. Finally, between R_7 and R_8 one has simultaneously the two chaotic attractors originated from the Feigenbaum cascades of such orbits. Summarising, we find the simulataneous existence of different periodic orbits and chaotic attractors. Hysteresis is a very common phenomenon in non linear systems and we do not comment any longer on it.

Transitions and crises of attractors.

An interesting feature of our system concerns the changes occurring in the geometry of various attractors. We start with the transition from quasi-periodic regime, associated to the invariant torus T_1 , to the chaotic regime CH_2 . This transition takes place in two steps: first, the motion becomes periodic, then chaos arises through a cascade of period doubling bifurcations. The first step occurs at the critical point R_1 where a pair of periodic orbits, $P_2^{1,2}$, are produced on the torus by a tangent bifurcation. Fig. 2 shows the attractors below and above R_1 . The birth of the P_2's on the torus is preceded by a series of frequency lockings with rotation number n/n+1, occurring at R = R_n

$$R_1-R_n = 1/n$$

Such a relation can be understood in connection with the intermittency developing near the bifurcation point R_1 (see for instance Ref. 2). Even though the quasi-periodic motion is destroyed at R_1, from a geometrical point of view the invariant two-torus persists beyond that point. In fact the unstable invariant manifolds $W^u(P_2')$ of the hyperbolic orbit P_2 form an invariant two-torus, while the eigenvectors of the linearized motion around the stable orbit P_2 , take on , first, real positive values and then complex conjugate ones. In the former case (positive eigenvectors), the manifolds $W^u(P_2')$ join smoothly on the P_2's, while in the latter (complex conjugate ones) they start wrapping around the P_2's, but still give rise to an invariant surface topologically equivalent to a torus. This is not true any more when the eigenvalues, become already equal and negative, remain less than zero, one with decreasing, and the other with increasing absolute value . Fig. 3 shows the $W^u(P_2')$ just below and above the parameter value R where $\lambda_1 = \lambda_2 = -.020$. The break up of the invariant torus seems clearly indicated. The second step of the overall transition torus-chaos takes place with the cascade of period doubling bifurcations starting from the P_2's.

Fig. 4 refers to an attractor crisis (see Ref. 3) taking place at R_4.

The two chaotic attractors, originated from the Feigenbaum cascades of the two P_2's (each of them made of two pieces in the stroboscopic images), join together when the corresponding hyperbolic orbits P_2' collide with them. The chaotic attractors, sitting before the collision on the unstable manifolds of the P_2's, extend, after the collision, also onto the unstable manifolds of the P_2 's. Such phenomenon is referred to as an "interior crisis" in the literature.

The chaotic attractor CH_2 evolves smoothly from $R = R_4$ to $R = R_6$, where it suddenly disappears ,colliding with the hyperbolic orbit P_3 (see Fig. 5a). As a result, P_3 survives as unique attractor. Such an event is usually indicated as "external" crisis.

An interior crisis, analogous to that described at R_4, occurs also at R_8. The two chaotic attractors $(CH_3)^{1,2}$, coming from the Feigenbaum cascades of the asymmetric P_{3a}'s collide with the symmetric P_{3s} , become hyperbolic after the symmetry-breaking bifurcation, and form the single attractor denoted with CH_3 . Note that the crisis restores, in this case, the symmetry of the attractor: the $(CH_3)^{1,2}$ are conjugate of each other, while CH_3 is itself symmetric. In Fig. 6, where the process above described is illustrated, each attractor appears made of three pieces.

CH_3 evolves smoothly between R_8 and R_9 . Its sudden change at R_9 is due to its collision with the hyperbolic orbit born together with P_{3s} . Fig. 7a shows the approaching collision, while Fig. 7b shows the new attractor CH produced by it.

A last discontinuous change in the attractor structure occurs at R_{11}, where a chaos-torus transition, similar to that already described, occurs. However, the smallness of the parameter region where it develops and the long duration of the period (13T) make a detailed analysis unfeasible.

Conclusions

We have given evidence for a break-up of an invariant torus in a transition from quasi periodic motion to chaos with an intermediate periodic regime. This route to chaos has been already theoretically described (4,5). Moreover, a first step has been made towards a systematic description of this model, with particular attention on the role played by the periodic orbits in connection with both quasi-periodic regimes (frequency lockings, break up of tori) and chaotic regimes (sudden merging or disappearence through collision with hyperbolic orbits).

References

1 R. Shaw, Z. Naturforsh. 36A,80 (1981).
2 K. Kaneko, Prog. Theor. Phys. 68, 669 (1982).
3 C. Grebogi, E. Ott, J.A. Yorke, Phys. Rew. Lett. 48, 1507 (1982).
4 S. Ostlund, D. Rand, J. Sethna, E. Siggia, Physica 8D, 303 (1983).
5 R.S. MacKay, C. Tresser, IHES preprint 1984

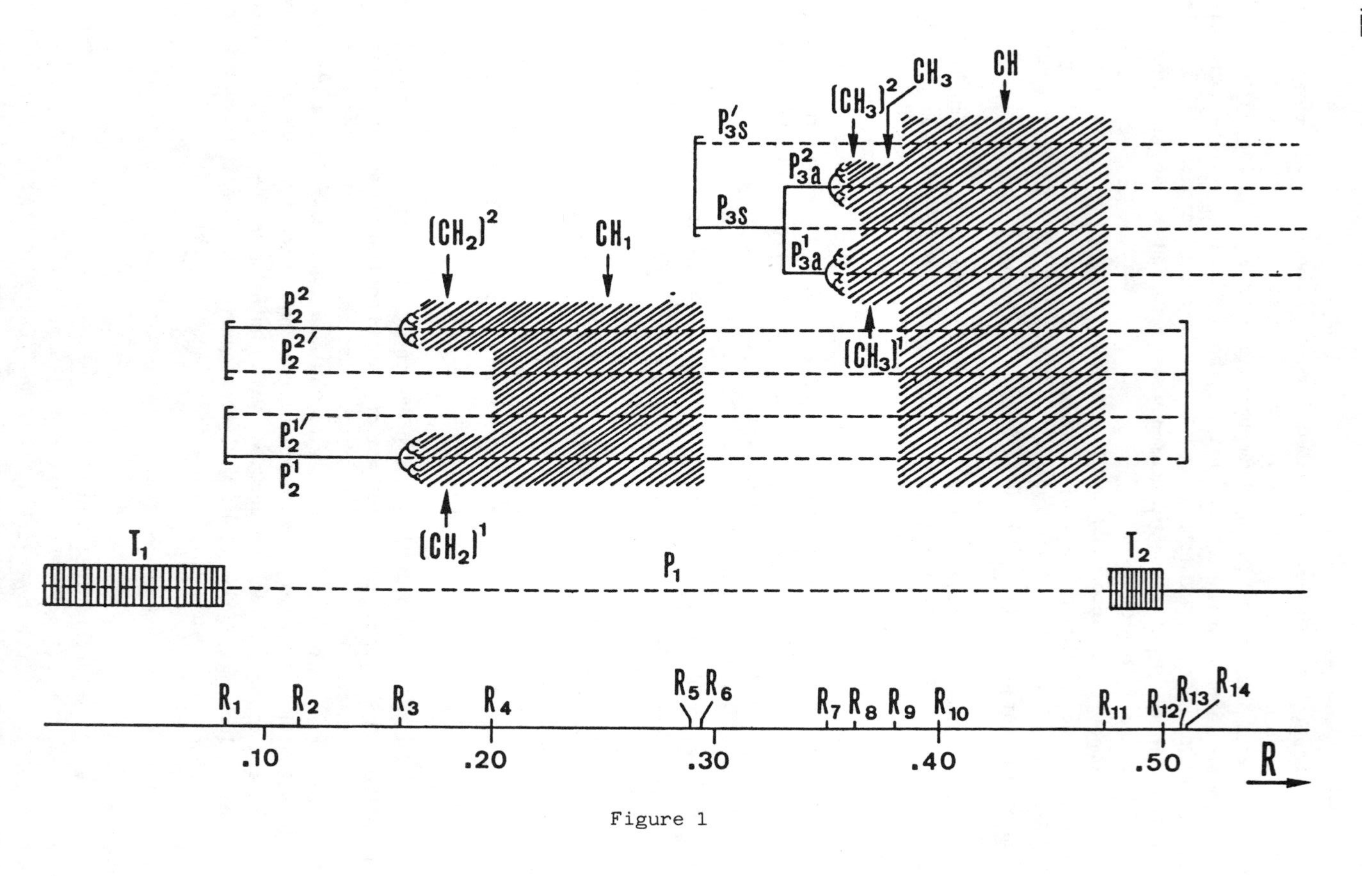

CH
CH_3
$(CH_3)^2$
P'_{3S}
P^2_{3a}
P_{3S}
P^1_{3a}
$(CH_3)^1$
$(CH_2)^2$
CH_1
P^2_2
$P^{2'}_2$
$P^{1'}_2$
P^1_2
$(CH_2)^1$
T_1
P_1
T_2
R_1
R_2
R_3
R_4
R_5 R_6
R_7 R_8 R_9 R_{10}
R_{11} R_{12} R_{13} R_{14}
.10
.20
.30
.40
.50
R

Figure 1

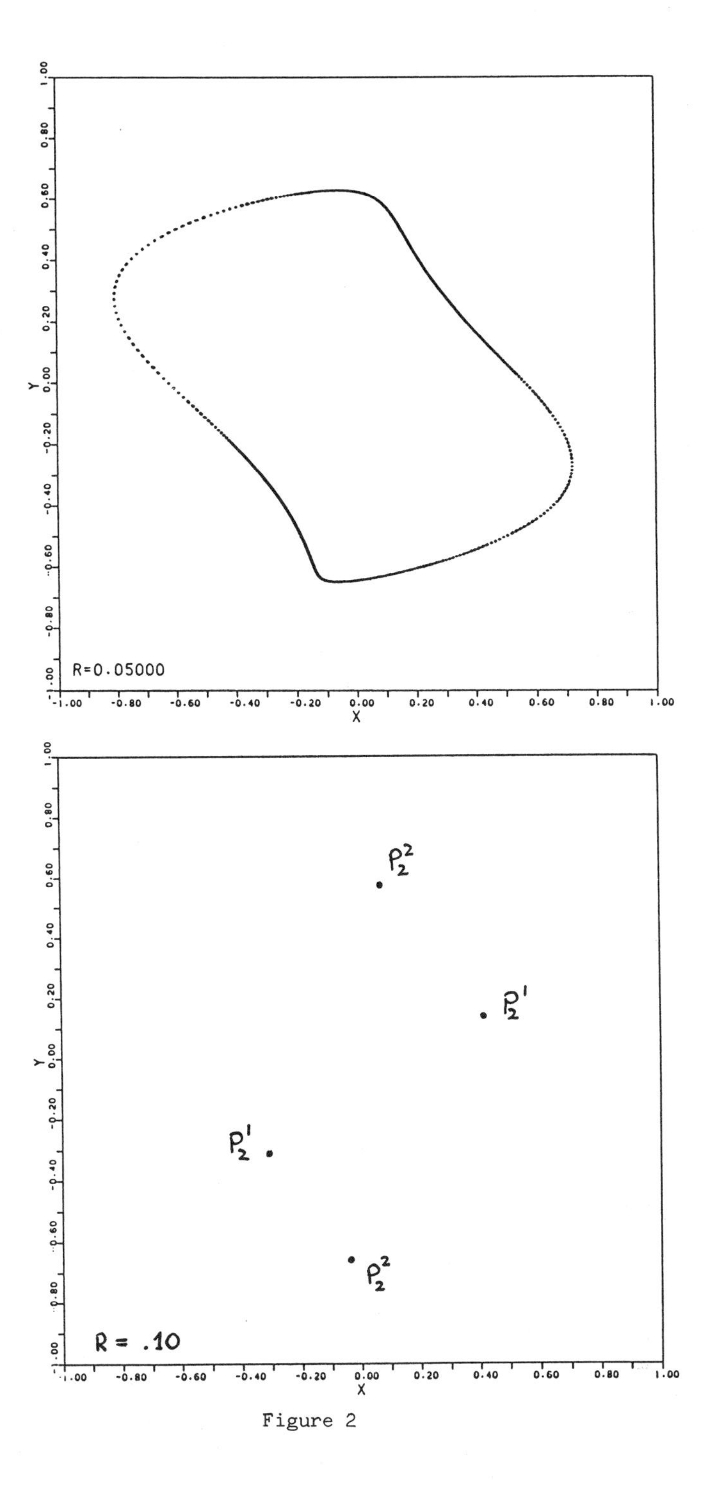
R=0.05000
Y
X
0.00
0.20
0.40
0.60
0.80
1.00
-0.20
-0.40
-0.60
-0.80
-1.00
P_2^2
P_2^1
P_2^1
P_2^2
R = .10

Figure 2

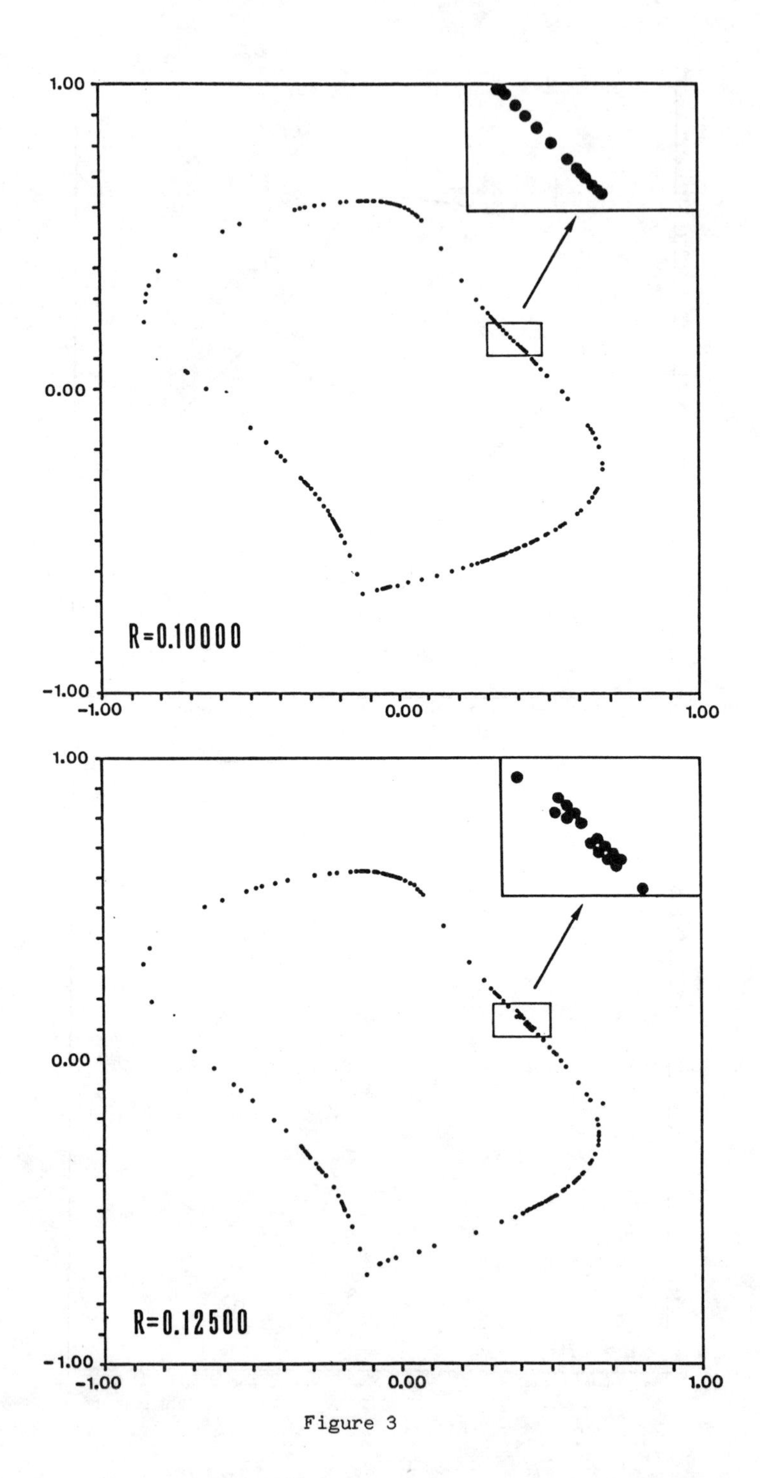
1.00
0.00
-1.00
-1.00
0.00
1.00
R=0.10000
1.00
0.00
-1.00
-1.00
0.00
1.00
R=0.12500

Figure 3

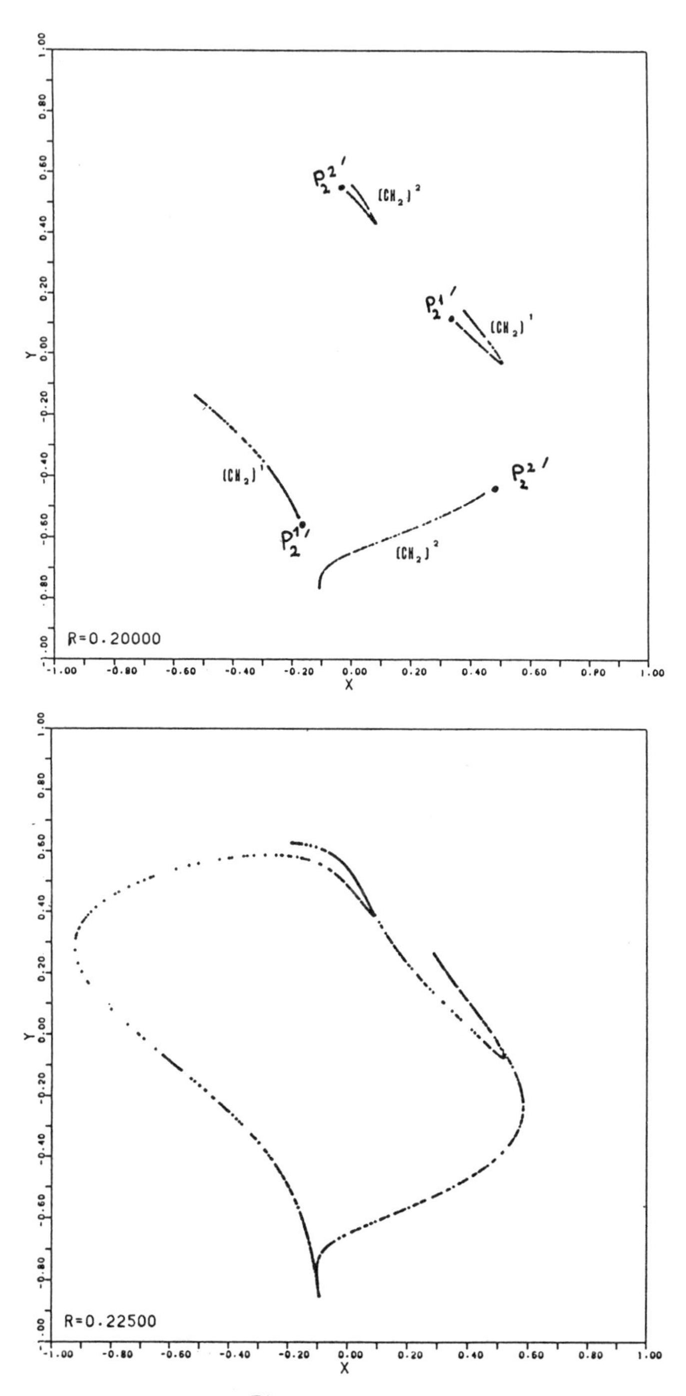
$P_2^{2\prime}$
$(CH_2)^2$
$P_2^{1\prime}$
$(CH_2)^1$
$(CH_2)^1$
$P_2^{2\prime}$
$P_2^{1\prime}$
$(CH_2)^2$
R=0.20000
Y
X
R=0.22500
Y
X

Figure 4

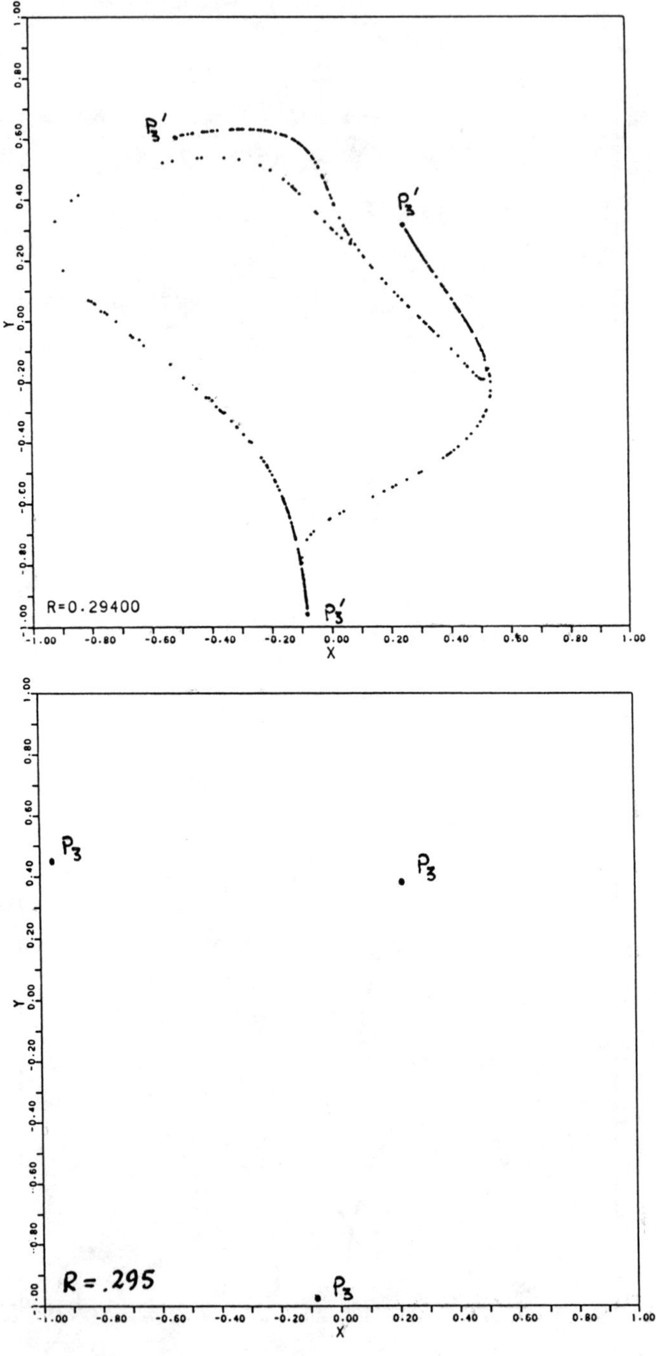

Figure 5

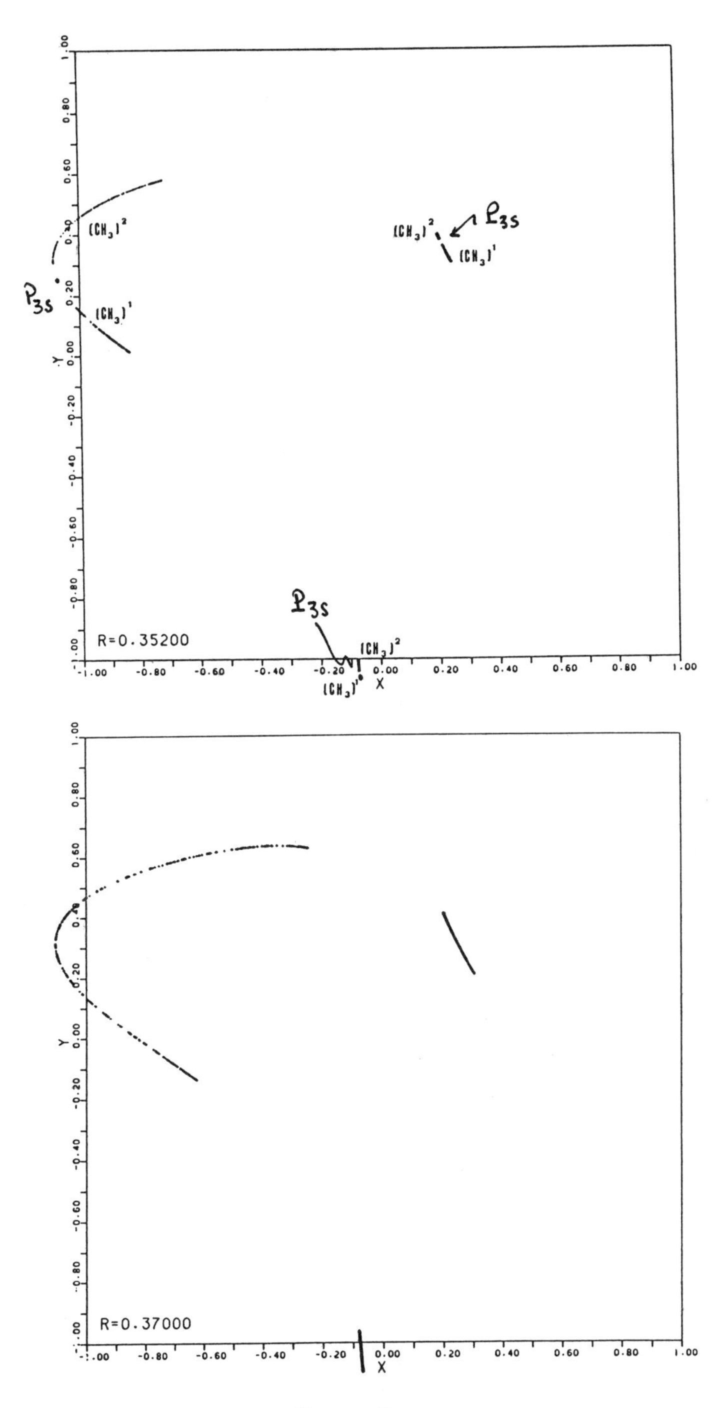

Figure 6

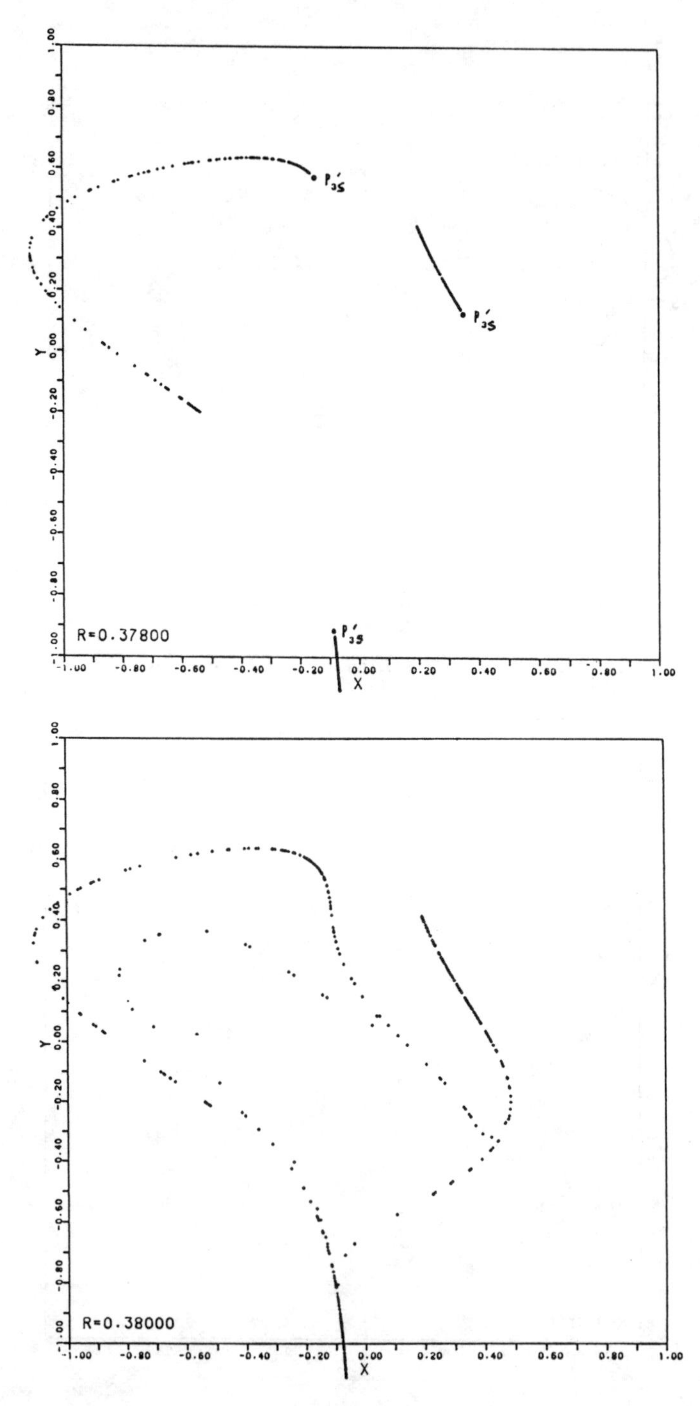

Figure 7

BIFURCATION PHENOMENA IN THE HÉNON MAPPING

Gianni Cocconi and Valter Franceschini

Dipartimento di Matematica Pura ed Applicata
Università di Modena, via Campi 213/B, 41100 Modena, Italy

Abstract : The Hénon mapping is investigated in a small region of the a,b parameter plane. Interesting bifurcation phenomena associated with the cycles of period seven are found, including the occurrence of bubbles due to remerging Feigenbaum trees. Such phenomena provide further evidence of how intriguing the behavior of a dissipative dynamical system may be under variation of more than one parameter.

The two-dimensional quadratic mapping T of equations

$$T(x,y) = (1+y-ax^2, bx),$$

represents one of the simplest dynamical systems that exhibit chaotic behavior for large sets of parameter values and initial conditions. Introduced by Hénon in 1976 [1], this mapping certainly attains the object of providing a model which reproduces the essential features of the Lorenz system [2], but much easier to be investigated by both theoretical and numerical tools.

Following Hénon, Feit [3], Curry [4] and Simo' [5] made numerical studies keeping b fixed and equal to 0.3; particular attention was devoted to the analysis of the strange attractor present for a=1.4. Hitzl [6] determined, in the a,b plane, the regions where escape to infinity occurs. More recently Hitzl and Zele [7] made a systematic investigation of the stable periodic orbits, of period 1 through 6, over the extended region $-2\leqslant a\leqslant 6$, $-1\leqslant b\leqslant 1$.

Among rigorous results, the existence of a transversal homoclinic orbit was proven by Marotto [8] for a>1.55 and b small enough, by Misiurewicz and Szewc [9] and Franceschini and Russo [10] in the case a=1.4 and b=0.3. An analogous result was obtained by Tresser et al. [11], who proved the existence of a topological horseshoe for parameter values associated with a numerically observed strange attractor.

Here we report on the results of a numerical investigation, based on bifurcation theory (see, for instance, Iooss and Joseph [12]), about the behavior of the Hénon mapping in a small region of the a,b plane which is included in the rectangle $1.22\leqslant a\leqslant 1.33$, $0.290\leqslant b\leqslant 0.308$. At first our object was a description, more detailed than the one known to us from the literature, of the phenomenology concerning the two cycles γ_1 and γ_2 of period seven which are present, for b=0.3, at a=1.24 and a=1.30 respectively. The two cycles may be of some interest because they lead to the formation of two strange attractors, the former occurring between the cycles, the latter being the same attractor found by Hénon at a=1.40. Moreover, we wanted to investigate the possibility that the ranges of stability associated with the two cycles, which are distinct for b=0.3, could be superimposed for a different value of b. After a preliminary investigation we became aware that many delicate bifurcation phenomena can occur in very narrow parameter ranges as both a and b are varied simultaneously. Hence, shedding light upon some of these phenomena became our main object.

The main tool to search for and follow periodic orbits is the Newton method. The iterative procedure for a cycle of period N, obtainable by solving a fixed point equation for T^N, is

$$[DT^N(P_k)-I]\Delta P_k = P_k - T^N(P_k), \; k=0,1,\ldots,$$

where $P_k=(x_k,y_k)$, $\Delta P_k=P_{k+1}-P_k$, I is the unit matrix and $DT^N(P_k)$ is the Jacobian of T^N at P_k. The iteration, which is started from an initial point P_o, is stopped when ΔP_k is as small as desired. $DT^N(P_k)$ can be easily computed recalling that

$$DT^N(P_k) = DT(T^{N-1}(P_k))\circ\cdots\circ DT(T(P_k))\circ DT(P_k).$$

Table I collects the critical values of the parameter a that summarize the behavior of γ_1 and γ_2. The value a_j^o, j=1,2, corresponds to the saddle-node bifurcation that originates γ_j. The value a_j^1 corresponds to the period doubling bifurcation that makes γ_j become unstable giving rise to a stable cycle γ_j^1 of period fourteen. In general, a_j^k is associated with the period doubling bifurcation from γ_j^{k-1} to γ_j^k, where γ_j^k is the cycle of period $7\cdot 2^k$ ($\gamma_j^o \equiv \gamma_j$). We were able to compute up to ten period doublings. The absolute error in the numerical values reported in the table is less than one unit in the rightmost digit.

Table I. Bifurcation points and Feigenbaum ratios associated with the sequences $\{\gamma_1^k\}$ and $\{\gamma_2^k\}$ for b=0.3

k	a_1^k	δ_1^k	a_2^k	δ_2^k
0	1.226617378	-	1.299116053	-
1	1.2541834643	-	1.3038207583	-
2	1.26001517267	4.72	1.30582162236	2.35
3	1.261428910473	4.13	1.30627536144	4.41
4	1.261738716104	4.56	1.306374024403	4.60
5	1.2618048038455	4.688	1.306395216870	4.656
6	1.2618189529498	4.6708	1.3063997586742	4.666
7	1.2618219827860	4.6699	1.3064007314901	4.6686
8	1.26182263166851	4.66931	1.30640093984955	4.66906
9	1.26182277063840	4.66923	1.30640098447403	4.669176
10	1.26182280040146	4.669208	1.30640099403125	4.669189

Together with the a_j^k's, the table includes the ratios

$$\delta_j^k = (a_j^{k-1}-a_j^{k-2})/(a_j^k-a_j^{k-1})$$

which express the bifurcation rate. They are clearly consistent with Feigenbaum theory [13] which states that, in any case of an infinite sequence of period doublings in a dissipative system,

$$\lim_{k\to\infty} \delta_j^k = \delta = 4.669201\ldots \ .$$

In comparison with analogous results (see, for instance, Derrida et al. [14]), the ratios δ_j^{10} represent better approximations of δ. For convenience, let us introduce the symbols a_j^∞ to represent the accumulation points of the sequences a_j^k, with the assumption that they are actually infinite.

The behavior of the sequences $\{\gamma_1^k\}$ and $\{\gamma_2^k\}$ in the a,b parameter space is displayed in Fig. 1. The picture represents the bifurcation curves $a_1^k(b)$ and $a_2^k(b)$ for k=0,1,2,3, with the addition of a curve $a_3^0(b)$ which corresponds to the appearance of an extra stable cycle γ_3. Ⓐ denotes a very small region where the curve $a_2^0(b)$ comes to an end for $b=b_2$ and the curve $a_3^0(b)$ begins for $b=b_1$. Ⓑ, Ⓒ, Ⓓ and Ⓔ correspond instead to places where we met with difficulties, particularly in determining the third period doubling. We shall examine in details which behavior occurs in the regions Ⓐ and Ⓓ.

Fig. 1 provides a clear idea of what happens as the parameter b is varied from the value 0.3. For increasing b the two ranges (a_j^0,a_j^∞) associated with the sequences of cycles $\{\gamma_j^k\}$, j=1,2, tend to separate more and more, while for decreasing b they tend to each other, getting partially superimposed before the end of $a_2^0(b)$.

To understand the behavior for $b<b_1$ a detailed analysis of the region Ⓐ was needed, and the result is Fig. 2. The picture, which consists of four "photograms" associated with decreasing values of b, is obtainable in the following way. Suppose that, for some pair (a,b) of parameter values, both γ_1 and γ_2 are present and close. It is straightforward to associate with each point P_1^k of γ_1, k=1,...,7, its neighbour P_2^k of γ_2. If there are other cycles of period seven in a neighbourhood of γ_1 and γ_2, such an association is immediate. By choosing one k, no matter which one, and a coordinate, for instance x, the plot of this coordinate as a function of a, for the k-th points of all the cycles that are involved, displays the evolution of the situation for the fixed value of b.

Consider then Fig. 2. The first photogram, which is relative to b=0.29427, shows the situation for $b>b_1$. The saddle-node bifurcation that originates γ_2 and its unstable twin $\bar{\gamma}_2$ at a_2 takes place without anything happening to γ_1. The second photogram corresponds to b=0.29423 and shows something new. Another stable cycle γ_3 arises in consequence of a saddle-node bifurcation which occurs for $a=a_3$. Slightly after its appearance, its unstable twin $\bar{\gamma}_3$ goes to meet γ_1 disappearing with it for $a=a_1^*$, where $a_3<a_1^*<a_2$. Nothing changes in the behavior of γ_2. The third photogram, obtained for b=0.294207, is essentially the same as the previous one, with

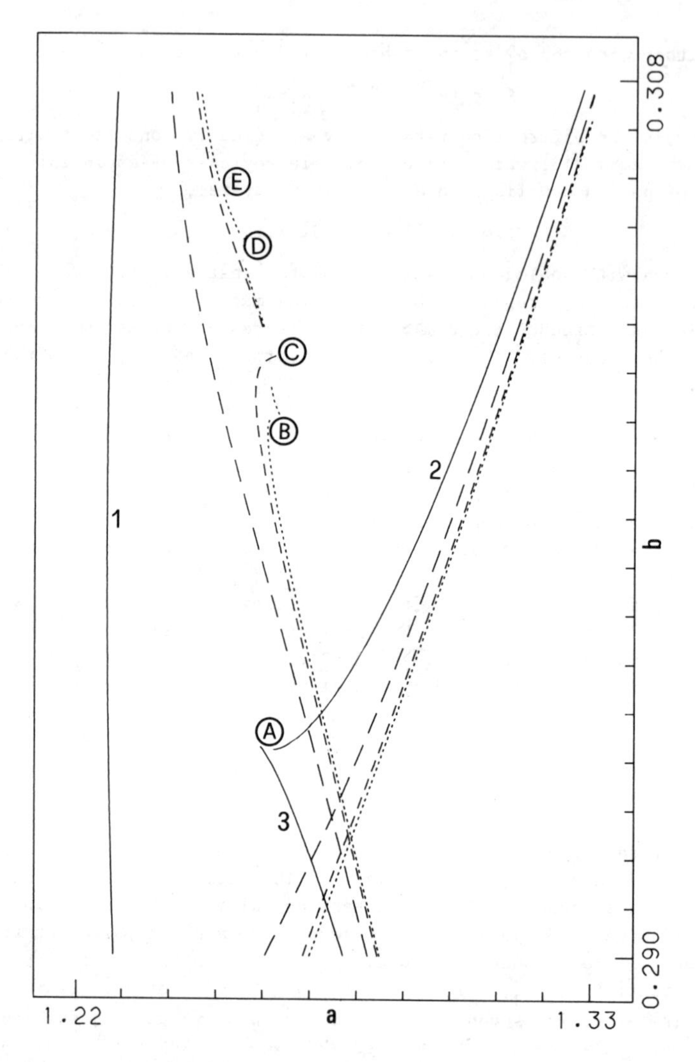

Fig.1. Bifurcation curves in the a,b parameter plane associated with cycles of period seven or multiple in the Hénon mapping. Solid lines represent saddle-node bifurcations, while long-dashed, short-dashed and dotted lines represent period doubling bifurcations, from period 7 to 14, from 14 to 28 and from 28 to 56 respectively. The integers 1, 2 and 3 refer to the cycles γ_1, γ_2 and γ_3. Encircled capital letters denote regions to be investigated more deeply.

the only difference that now a_1^* and a_2 have got much closer. The last photogram shows, for b=0.294206, a final different phenomenology. The curves associated with γ_1 and γ_2 have joined for $b=b_2$ and the two saddle -node bifurcations at a_1^* and a_2 do not occur any more.

A numerical definition of b_1 and b_2 turns out from the four photograms of Fig. 2: $0.29423<b_1<0.29427$, $0.294206<b_2<0.294207$. Moreover, an interesting remark can be made in the case $b=b_2$ about the bifurcation that occurs at $a_1^*=a_2$. Rather than two joined saddle-node bifurcations, it is an exchange of stability between γ_1 and γ_3. Trying to summarize in a few words what happens in the region Ⓐ, one can say that γ_1 first gives up its behavior to a new cycle γ_3 and then takes possession of the behavior of γ_2 by joining it.

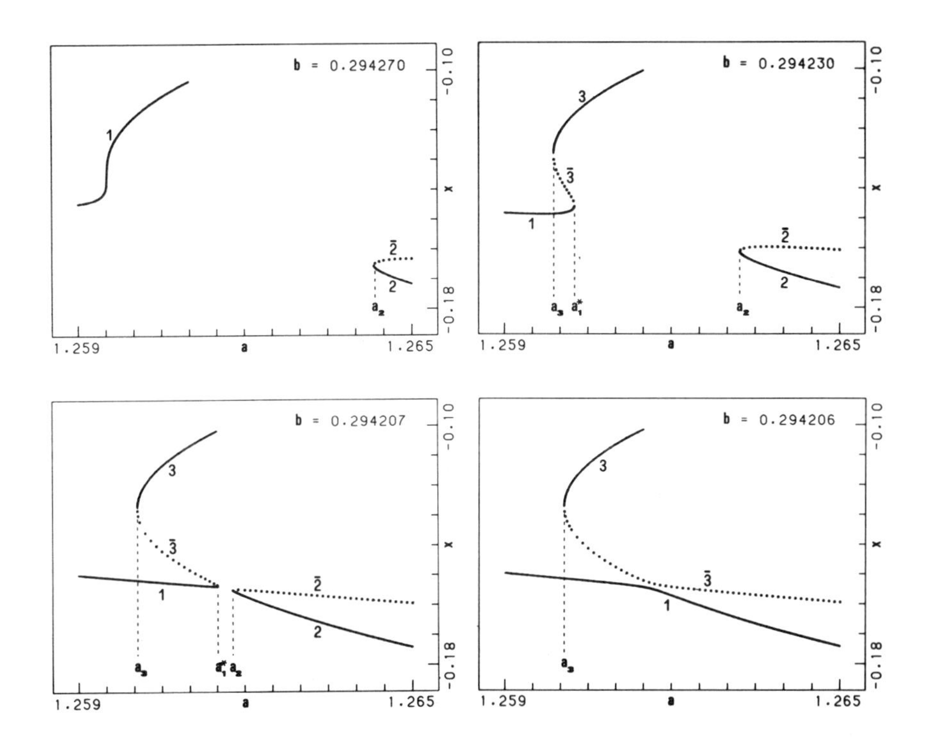

Fig.2. Diagrams in the a,x plane which explain the behavior in the region Ⓐ. Solid lines correspond to stable cycles, while dotted lines correspond to unstable cycles. There are shown: for b=0.294270 γ_1, γ_2 and $\bar{\gamma}_2$; for b=0.294230 and b=0.294207 γ_1, γ_2, $\bar{\gamma}_2$, γ_3 and $\bar{\gamma}_3$; for b=0.294206 γ_1, γ_3 and $\bar{\gamma}_3$.

The phenomenology that takes place in the region (D) of Fig. 1 is really impressive. To illustrate it, we proceed as before by providing a sequence of six photograms (Fig. 3) corresponding to different fixed values of b, also here taken in decreasing order. Each photogram is obtained by following the cycle γ_1^2 as a is increased. Analogously to Fig. 2, we plot the abscissa of one point, say P_k, among the 28 points of γ_1^2. To represent a cycle bifurcated from γ_1^2 via period doubling, we plot the abscissas of both the points originated from P_k. In the case of a further doubling, we plot the four points that correspond to P_k. If other cycles exist in a neighbourhood of γ_1^2, we represent the points that are associable with P_k by closeness.

Before explaining Fig. 3, it must be said that an involved sequence of period doubling and saddle-node bifurcations makes it quite meaningless to maintain the notations used up to now for the cycles. So, we shall refer to them simply by using their period.

The first photogram of Fig. 3, relative to b=0.3045, shows a usual situation, with the cycle 28 bifurcating into the cycle 56 (at a_1^3). Successive period doublings have been verified, but they are not represented.

The second photogram displays new behaviors which already occur for b=0.3044. First, the cycle 28 becomes unstable at a' bifurcating into a cycle 56 and rapidly recovers stability at a" because of a reverse period doubling bifurcation, which is due to the return onto it of the same period 56. Second, the cycle 28 disappears in consequence of a saddle-node bifurcation which leads it to collapse with a nearby unstable cycle of the same period. The latter was just born together with another stable cycle 28 which, as a increases, gives rise again to a sequence, presumably infinite, of period doublings. This second part of phenomenology will not change throughout the remaining photograms of Fig. 3.

Restricting our consideration to what happens between a' and a", we observe that for b=0.30432 (see the third photogram of the figure) the behavior associated with the cycle 56 is perfectly analogous to the one of the cycle 28. In fact, first it becomes unstable bifurcating into a stable cycle 112, then it becomes again stable thanks to the remerging of the cycle 112.

The fourth photogram, which is associated with b=0.30425, shows a further change of the behavior between a' and a". Also the period 112 loses and recovers stability as a consequence of two period doubling bifurcations, the first one for increasing a and the second one for decreasing a. But differently from the two previous cases, now the sequence of period doublings appear to be infinite on both sides. This is supported by a check, made by computing the Liapunov exponents for a=1.2565, of the presence of a strange attractor in between.

The fifth photogram, which corresponds to b=0.30415, exhibits a behavior much more complicated. Without entering into details, we observe that new pairs of sequences of presumably infinitely many period doublings take place, presumably with a strange attractor in between. Such pairs are separated from each other by pairs of saddle-node bifurcations associated with cycles of period 56 or 112. In these cases an unstable cycle

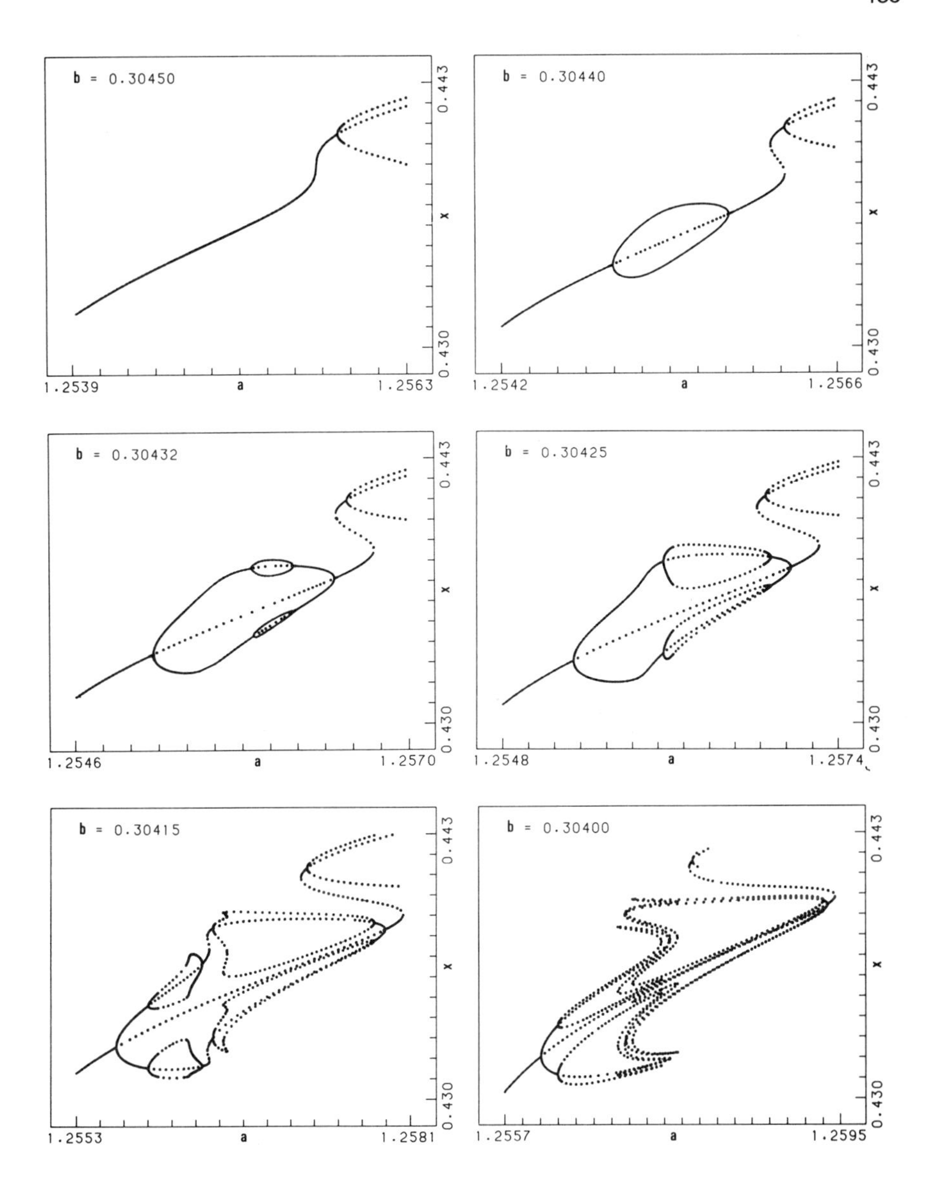

Fig.3. Diagrams in the a,x plane which show the behavior associated with the cycle γ_1^2, of period 28, in the region Ⓓ. Solid lines and dotted lines represent stable cycles and unstable cycles respectively.

makes first a new stable cycle arise and then a pre-existent one disappear.

As b decreases further, the number of these bifurcation phenomena between a' and a" increases, while the intervals of a in which they occur become narrower and narrower. The sixth and last photogram, relative to b=0.30400, besides suggesting that now the ranges of chaotic behavior cover almost completely the interval (a',a"), provides evidence of how intricated things are becoming. We did not investigate the way that leads again to a phenomenology like that of the first photogram of Fig. 3 as b is furtherly decreased.

The presence in dynamical systems of "bubbles" due to remerging period doubling sequences is not uncommon. Among others it was found by Contoupulos [15], Oppo and Politi [16], Bier and Bountis [17]. In [17] the occurrence of high-order bubbles is explained with the presence of a symmetry in the equations as well as with the occurrence of a primary bubble of period 2. Our results on the Hénon mapping indicate that these conditions are not essential in order that such phenomena take place.

Coming back to one of our initial questions, we remark that, as b decreases and the a-ranges associated with the cycles γ_1 and γ_2 tend to each other, many interesting phenomena of hysteresis, i.e. coexistence of different attractors, occur. Among such phenomena, for b=0.295 we observed that the strange attractor originated by the sequence $\{\gamma_1^k\}$ and the stable cycle γ_2 are present at the same time. They coexist up to the appearance, for a value of a just below 1.2747, of a stable cycle of period 21 which takes the place of the strange attractor. A sequence of period doublings of this cycle, accumulating around a=1.27497, reinstates the strange attractor. As a is increased further, a crisis, occurring for a between 1.2772 and 1.2775 owing to the collision with the unstable cycle $\bar{\gamma}_2$, causes its final destruction (see Grebogi et al. [18]).

To conclude, let us make some final comments. We reported the results of a detailed investigation of the Hénon mapping in a small region of the a,b parameter plane. Dealing only with cycles of period seven and derived attractors, we showed that very involved sequences of bifurcations may occur as both a and b are varied. In particular we demonstrated, at our knowledge first in the case of the Hénon mapping, the existence of remerging Feigenbaum sequences, which were found to take place in quite narrow parameter ranges. Besides adducing a little contribution to the collection of phenomenological data for the Hénon mapping, we hope we have provided further stimulating evidence of how intriguing the behavior of a dissipative dynamical system may be when one looks carefully at the details.

Acknowledgements. The Centro di Calcolo of the University of Modena is acknowledged for providing computer facilities and financial support.

References

[1] M. Hénon, Commun. Math. Phys., 50, 69 (1976).
[2] E. N. Lorenz, J. Atmo. Sci., 20, 130 (1963).
[3] S. D. Feit, Commun. Math. Phys., 61, 249 (1978).
[4] J. H. Curry, Commun. Math. Phys., 68, 129 (1979).
[5] C. Simo', J. Stat. Phys., 21, 465 (1979).
[6] D. L. Hitzl, Physica, 2D, 370 (1981).
[7] D. L. Hitzl and F. Zele, preprint (1984).
[8] F. R. Marotto, Commun. Math. Phys., 68, 187 (1979).
[9] M. Misiurewicz and B. Szewc, Commun. Math. Phys., 75, 285 (1980).
[10] V. Franceschini and L. Russo, J. Stat. Phys., 25, 757 (1981).
[11] C. Tresser, P. Coullet and A. Arneodo, J. Phys. A : Math. Gen., 13, L123 (1980).
[12] G. Iooss and D. D. Joseph, Elementary Stability and Bifurcation Theory, Springer-Verlag, New York, Heidelberg, Berlin (1980).
[13] M. J. Feigenbaum, J. Stat. Phys., 19, 25 (1978).
[14] B. Derrida, A. Gervois and Y. Pomeau, J. Phys. , A12, 269 (1979).
[15] G. Contopoulos, Lett. Nuovo Cimento, 37, 149 (1983).
[16] G. L. Oppo and A. Politi, Phys. Rev., 30A, 435 (1984).
[17] M. Bier and T. C. Bountis, Phys. Lett., 104A, 239 (1984).
[18] C. Grebogi, E. Ott and J. A. Yorke, Phys. Rev. Lett., 48, 1507 (1982).

QUALITATIVE ANALYSIS OF THE LORENZ EQUATIONS

Massimo Miari

Dipartimento di Fisica dell'Università di Milano

Via Celoria 16 - MILANO (Italy)

Abstract. In the present paper we reduce the Lorenz equations to a well known dynamical system on a two-dimensional center manifold. This is a classification result, which provides a syntetic qualitative description of some relevant aspects of the phase portraits and of the bifurcations in the original system. The theoretical justifications for our work, in the light of bifurcation theory for vector fields, and the consequences of such a result are also emphasized.

1. Context

Up to this time the Lorenz equations

$$\begin{aligned} \dot{X} &= -\sigma(X-Y) \\ \dot{Y} &= rX - Y - XZ \\ \dot{Z} &= -bZ + XY \quad , \end{aligned} \tag{1}$$

where $(X,Y,Z) \in \mathrm{R}^3$, and σ, r, b are positive parameters, altough thoroughly studied, have been seen as a strange "pathological object", unique in its kind, born by chance in a skilful reduction of the Boussinesq-Oberbeck equations for a two-dimensional convection problem [1]. With this picture in mind, one tried to recognize some known features, namely the attractors for the system and the codimension-one bifurcations arising in connection with the loss of stability of the fixed points. The Lorenz system has so been dissected, varying the parameter r while keeping fixed at suitable values the parameters σ and b, in opens of stability and close sets of bifurcation. For example, system (1) was reduced, in the neighborhood of $r = 1$, using local codimension-one bifurcation and the center manifold method, to a one-dimensional system describing a supercritical pitchfork bifurcation [2], while Marsden and McCraken [3] have shown that at a certain value $r = r_H$ one has a subcritical Hopf bifurcation. It is of interest to notice that these two results are to date the only analytical ones; for the remainder, the global codimension-one bifurcations, occurring when a non-transverse connection between stable and unstable manifolds of a fixed point or of a closed orbit appears, have only been conjectured but never proved analytically. The same has happened for the description of the bifurcations of the strange attrctor at high values of the parameter r, as for example the inverse cascade of subharmonic bifurcations [4]; one recognizes in the numerical integration of the Lorenz system a codimension-one bifurcation, then in the neighborhood of the bifurcation value of the parameter r the system is thought of as a single one, disjoined from the systems corresponding to other values of r.

The reason for such a way of proceeding may be envisaged in the main interest in a dynamical system, like the Lorenz one, which could model the chaos and the onset of turbulence in a finite macroscopic system. Most of the efforts were addressed to study the details of the strange attractor and the degree of chaos of the motion on it, without studying how such a strange attractor rised with its shape and if its birth was unavoidable.

On the contrary, we think that it is time now that the research be directed towards the problems, less spectacular but equally interesting, of morphogenesis, i.e. the study of the birth and the changes of the shape of an object. For what concerns the Lorenz equations, this aim is traduced in the attempt to reduce it to a system in normal form, for which all the bifurcations occurring in the Lorenz system, or at least the two-dimensional ones, are justified and unavoidable, i.e. the sequence of these bifurcations is generic and characteristic of an equivalence class of dynamical systems structurally stable.

Indeed, the mathematical tools at our disposal are effective enough to pursue such a result. First of all, center manifold theory permits us to reduce the number of degrees of freedom of the system, by considering only the motion on the local manifold on which the modifications in the phase portrait of the system occur. If we are dealing with bifurcations from equilibria, such a center manifold W^c will be tangent to the center eigenspace E^c associated to those eigenvalues of the Jacobian matrix of the system having zero real part. Moreover, if there is no unstable manifold, then the local attractivity of W^c guarantees that the system reduced on W^c determines the true asymptotic behavior of all the solution curves. Since the center manifold W^c is invariant under the flow, the reduced essential system can be written by replacing the local coordinates in the center manifold by the coordinates in the center eigenspace, i.e. projecting the flow on W^c onto E^c. A theorem due to Carr [5] shows then that we can approximate the function, whose local graph determines the center manifold in the neighborhood of the non-hiperbolic fixed point, to any degree of accuracy by a polynomial approximation, under the hypotesis that the center manifold is sufficiently smooth on some neighborhood of the critical point. We will see later that the neighborhood on which the smoothness hypotesis holds may be very narrow in the parameter space.

To enrich the local analysis and choose the suitable center manifold, we have to consider the results of the qualitative theory of differential equations for codimension-two bifurcations, i.e. for the loss of stability of equilibria depending on two parameters. Indeed, all such bifurcations are classified and, for some of these, the modifications of the phase portrait as the parameters vary are known [2,6]. Rather surprisingly these modifications include also codimension-one global bifurcations, as for example the onset of a homoclinic connection or the birth of a couple of periodic orbits of opposite stability. For these codimension-two bifurcations the unfolding, i.e. the vector field in normal form, is also known; general theorems [6] finally state that the two-parameters family of vector fields is stucturally stable under small perturbations of the vector field and of the parameters. This means that the codimension-two bifurcation is unavoidable for the family when both parameters are varied.

The final aim of our work is the attempt to reduce, on the corresponding center manifold, the Lorenz system to one of the normal forms for a codimension-two bifurcation, in order to have a syntetical and analytical understanding of all the two-dimensional codimension-one bifurcation observed in the Lorenz system and to justify the three-dimensional ones. The reduction to such a well known dynamical system and the structural stability of the vector field in normal form obtained by a polynomial appoximation of the center manifold will insert the Lorenz system in a class of topological equivalence of systems which suffer the same codimension-two singularity and are invariant under the action of the same symmetry group. These are very common conditions, so that the number of objects in such a class is considerably high: among these we recall for example the systems obtained by Arneodo et al.[7], Coullet et al.[8] or Holmes [9] by reducing the equations of the hydrodynamics in the case of two or three competing instabilities. This loss of singleness suffered by the Lorenz system is however repaied by the increased knowledge about it and about the way a strange attractor may arise and take its own shape.

In the present paper I give a short review of the results I obtained recently [10] along such lines.

2. Conjecture and proof

In order to pursue the aim expressed in the previous section, we have to recall some of the properties of the Lorenz system and the bifurcations observed by varying the parameter r. First of all, the system (1) is invariant under the tranformation $(X, Y, Z) \rightarrow (-X, -Y, Z)$, i.e. under rotations through π around the Z axis. It has then been proved the existence of a closed surface in the phase space, for any point of which the vector field is directed inward; system (1) is a dissipative one, so that the volume of the phase space contracts itself on zero-measure sets which determine the long term behaviour of the orbits of the system. We are so brought to look for the attractors of the system.

The fixed points are the origin $P_0 = (0,0,0)$, representing the conductive equilibrium, and two convective equilibria $P_{\pm} = (\pm\sqrt{b(r-1)}, \pm\sqrt{b(r-1)}, r-1)$. These symmetrical fixed points exist only for $r > 1$.

For what concerns the bifurcations observed in the Lorenz system by varying the parameter r, these are summarized in Fig.1. For $r \in (0,1)$ the origin P_0 is the only attractor for the flow; it is initially an attracting focus (the eigenvalues of the Jacobian matrix are a couple of complex conjugated ones with negative real part plus a real negative eigenvalue) then a sink. For $r = r_p = 1$ one has a supercritical pitchfork bifurcation: the conductive equilibrium becomes a saddle while two symmetrical sinks arise from it. At a numerically quoted value $r = r_0 \simeq 13.93$ for $\sigma = 10$, $b = \frac{8}{3}$ a double homoclinic connection between the stable and unstable manifold of the origin is supposed to give rise to a strange invariant set. After the two sinks have become foci, one has a double subcritical Hopf bifurcation at $r = r_H$. The well known strange attractor exists however before the convective equilibria $P_{\pm}$ loose their stability: it arises at $r = r_s$ after a double non-transverse connection between the unstable manifold of the origin and the stable manifold of the closed orbits which desappear in the Hopf bifurcation [2,11]. At high values of r, such strange attractor seems to desappear, after an inverse cascade of subharmonics bifurcations, until the only attractor for system (1) is a stable closed orbit encircling the three fixed points [4,12].

Let us now consider the dynamical system

$$\begin{aligned} \dot{x} &= y \\ \dot{y} &= \mu_1 x + \mu_2 y \pm x^3 \pm x^2 y \quad ; \end{aligned} \tag{2}$$

it is the normal form of a vector field depending on two parameters μ_1 and μ_2, invariant under rotations through π, such that for $\mu_1 = \mu_2 = 0$ the Jacobian matrix at the origin takes the form of a nilpotent Jordan block of order two, namely $\begin{pmatrix} 0 & 1 \\ 0 & 0 \end{pmatrix}$. This is the codimension-two singularity we consider hereafter, and system (2), to use Arnold's terminology [6], is the versal deformation of such singularity.

The modifications of the phase portrait as the parameters are varied are well known [2,6] and are illustrated in Fig.2 in the case when the coefficients of the cubic terms are both negative. As one can see, by varying the parameters in such a way as to follow the path marked by arrows in the figure, one recognizes the sequence of two-dimensional codimension-one bifurcations which occurs in the Lorenz system. One can remark that in fact the unstable closed orbit encircling the three fixed points has never been observed in the Lorenz system, while the stable one seems to exist only for very high values of the parameter r; a conjecture about this point will be given in the closing section.

Starting from this leading analogy, we want to reduce the Lorenz system on a suitable two-dimensional center manifold corresponding to the onset of the codimension-two bifurcation presented above. After the reduction we will have to verify that the signs of the coefficients of the cubic terms are the correct ones in order to have the wanted unfolding. Furthermore, having fixed a posteriori the historical values of σ and b, we want to show that a path as indicated in Fig.2 is effectively followed when r is varied, so that any codimension-one bifurcation is met, at values of r which are to be compared with those numerically estimated.

At first sight it would seem natural to operate such a reduction in a neighborhood of the origin P_0. In order to do that, it is an easy matter to verify, studying the characteristic polynomial for the Jacobian matrix of system (1) at the origin, that the critical values of the parameters for which one has a nilpotent Jordan block of order two are $\sigma = -1$ and $r = 1$. Obviously a negative value of σ is of no physical meaning, but we have to recall that the reduction on the center manifold holds in a neighborhood of the critical values of the parameters, as well as in a neighborhood of the critical fixed point.

After a linear change of variables in the X, Y plane, in order to put the Jacobian matrix in normal Jordan form, the reduced system at the third order is obtained by the standard method, described for example in the paper of Holmes [9]. The details of the computations, here and afterwards, may be found in my previous paper [10]. By applying the normal form theorem, in order to eliminate the non-resonant terms from the reduced system, only in phase space, one gets a system on the center manifold which is, by suitably rescaling the variables, of the wanted form, namely

$$\begin{aligned} \dot{x} &= y \\ \dot{y} &= \mu_1 x + \mu_2 y - x^3 - x^2 y \quad , \end{aligned} \tag{3}$$

with

$$\begin{aligned} \mu_1 &= \left(\frac{(b+2)}{b}\right)^2 \frac{(r-1)}{\sigma} \\ \mu_2 &= -\frac{(b+2)}{b}\frac{(\sigma+1)}{\sigma} \quad . \end{aligned} \tag{4}$$

The result cannot nevertheless be considered satisfactory because μ_2 does not depend on r; hence, if one fixes a positive value of σ and lets r vary, μ_2 is also fixed and negative and one follows a straight line parallel to the μ_1 axis in the parameter plane, meeting only the first trivial modifications observed in the Lorenz system.

The suspicion that this unsatisfactory result may be caused by the method of approximation of the center manifold, namely by the failing of the hypotesis of analyticity of it in a neighborhood of the critical point, led us to face the problem "from the other side", around one of the symmetrical convective equilibria.

As we have already recalled, at $r = r_H$ two of the eigenvalues of the Jacobian matrix at one of these fixed points become purely immaginary and one has a subcritical Hopf bifurcation. The secular equation corresponding to the characteristic polynomial gives indeed a pair of purely immaginary eigenvalues $\lambda_{1,2} = \pm i\omega$ for $r = r_H$, where

$$r_H - 1 = \frac{(\sigma+1)(\sigma+b+1)}{(\sigma-b-1)} , \tag{5}$$

with

$$\omega = \sqrt{\frac{2b\sigma(\sigma+1)}{(\sigma-b-1)}} , \tag{6}$$

which is real for $\sigma > b+1$. Moreover one has an eigenvalue $\lambda_3 = -(\sigma+b+1)$.

The idea consists now in reducing system (1) around P_+ and in a neighborhood of the critical value r_H to the normal form (3). Two remarks are nevertheless necessary. First of all, in the choice of P_+ we have lost the symmetry of the Lorenz system, so that we have now to consider the system which is the unfolding of the same codimension-two singularity but without symmetry. It is well known [2,6] that such a system is the following:

$$\begin{aligned} \dot{x} &= y \\ \dot{y} &= \mu_1 + \mu_2 y + a x^2 + b x y \quad . \end{aligned} \tag{7}$$

When the coefficients of the nonlinear terms have the same sign, we have the analog of system (3); the modifications of the phase portrait as the parameters μ_1, μ_2 are varied are shown in Fig.3. The second remark concerns the fact that system (7) is the unfolding of a codimension-two bifurcation, while we are able, from the study of the secular equation in P_+, to meet only the condition for a codimension-one bifurcation, namely the Hopf one. In order to have a pair of vanishing eigenvalues one has to require $\omega = 0$, and this is impossible for $\sigma > 0$, $b > 0$. Thus the reduction to system (7) will be possible only by imposing the condition that one of the two parameters does not vanish.

As usual we choose the linear change of variables which leads the Jacobian matrix calculated at $r = r_H$ to the form

$$J = \begin{pmatrix} 0 & 1 & 0 \\ -\omega^2 & 0 & 0 \\ 0 & 0 & \lambda_3 \end{pmatrix} . \tag{8}$$

The parameter r enters now only through the function $\mu(r)$ given by

$$\mu(r) = \sqrt{\frac{(r-1)}{(r_H-1)}} - 1 = \sqrt{\frac{(r-1)(\sigma-b-1)}{(\sigma+1)(\sigma+b+1)}} - 1 \quad . \tag{9}$$

We are now in position to apply the center manifold technique at $(\underline{x}, \mu) = (\underline{0}, 0)$. In the lowest approximation, which corresponds to projecting the transformed system on the center eigenspace, one obtains a reduced system which belongs to the wanted equivalence class, up to terms of order not less than three in the variables and in the parameter. In particular the coefficients of the nonlinear terms, namely

$$\begin{aligned} a = & -\frac{\sigma(\sigma+1)}{(\sigma-b-1)}[2b(\sigma+b+1)(2\sigma-b)+2(\sigma+1)(\sigma^2-b-1)+ \\ & +(\sigma+b+1)^2(\sigma-b-1)+2b\sigma(\sigma+1)] \\ b = & -2(\sigma+1)[(\sigma+b+1)+b(b+1)] \quad , \end{aligned} \tag{10}$$

are easily checked to have the same sign if $\sigma > b+1$.

From the study of the reduced system one sees that it always admits two fixed points, one of which is a saddle point and should correspond to the conductive solution of the original Lorenz system, while the second one is the point around which the reduction was made. At $\mu_1 = -\mu_2{}^2$ one has a subcritical Hopf bifurcation; by construction, this correspond to $\mu = 0$, but now we have also the information about the stability of the periodic orbit, a result which was already obtained by Marsden and McCraken [3]. Finally, on the line O in Fig.3 given by $\mu_1 = -\frac{49}{25}\mu_2{}^2 + O(\mu_2{}^3)$ one has a homoclinic orbit; this corresponds to $\mu \simeq -0.0168$ for $\sigma = 10$ and $b = \frac{8}{3}$. From the explicit expression of $\mu(r)$ one gets correspondingly $r \simeq 22.94$, which should be compared with the numerical value, $r \simeq 13.93$, found for the Lorenz system.

We can look now for the expression of the center manifold in the first non- trivial approximation, as the graph of a function of second order in the variables and in the parameter. One thus gets the reduced system at third order. In addition to the origin, corresponding to P_+, such a system has two non-symmetrical fixed points Q_0 and Q_-, which we identify, after an analysis of their linear stability, with P_0 and P_- respectively. One can now revert to the original aim of obtaining system (3): we perform first of all the translation which bring the conductive equilibrium P_0, now seen as Q_0, to the origin, taking care to neglect terms of order greater than three. The original symmetry which was present in the Lorenz system as well as in system (3) can be regained by a nonlinear change of variables which eliminates the terms, occurring in the reduced system, which are quadratic in the coordinates of the center eigenspace. In such a way one gets a system of the form (3), with the correct signs for the coefficients of the cubic terms, at least for $\sigma = 10$, $b = \frac{8}{3}$.

At variance with the result obtained around the conductive equilibrium, we can now recover all the codimension-one bifurcations occurring in the Lorenz system on a two dimensional manifold. The values μ_p and μ_o, for the pithfork and the homoclinic bifurcations respectively, are found to be $\mu_p \simeq -1$ and $\mu_o \simeq -0.036$, which correspond to $r_p \simeq 1$ and $r_o \simeq 22.1$. The last value is still a rough approximation of the numerically determined one 13.93 ; the improvement with respect to the result obtained in the center eigenspace approximation justifies however our hope for better approximations at higher orders.

3. Discussion

In the previous section we have let two questions unresolved. First of all we have to explain why neither the global bifurcation occurring in system (3) on the line C, where a couple of periodic orbits of opposite stability arises, nor such periodic orbits encircling the three fixed points, at least for low values of r, are seen in the Lorenz system. Moreover, it is not clear why the reduction around the conductive equilibrium does not provide the full correct result.

We make the following conjecture: if the two-dimensional center manifold presents an angle along a direction containing the origin of the phase space for certain values of the parameters, as may actually be the case when the real negative eigenvalue $\lambda_2 = -\frac{\sigma+1}{2} - \sqrt{\frac{(\sigma+1)^2}{4} + 4\sigma(r-1)}$, corresponding to the eigenspace normal to the plane of the picture in Fig.1, becomes very large in modulus, then the polynomial approximation of the center manifold does not hold, due to failure of the smoothness condition. On the contrary, when one makes the reduction around one of the convective equilibria, the angle at the origin is approximated by a smooth surface and the resulting system gives a good description of the original one.

It should be clear that any periodic orbit encircling the three fixed points has to cross the edge of the angle, where the smoothness condition fails, so that the orbits should somehow break there and disappear. At high values of the parameter r the folded manifold may become smooth again so that the stable periodic orbit, having regained its differentiable character, should be the only attactor for the system.

In fig.4 we give an illustation of these ideas. In the left column the phase portraits of system (3) in four different regions of structural stability in the parameter plane are shown. The dashed line, marking the line of folding of the two-dimensional center manifold, has been chosen tangent to the unstable manifold of the origin, to assure that it is not folded and preserves its differentiable character. The appearence of the corresponding phase portraits after the folding is shown in the right column. One can remark that the motion appears to be in fair agreement with the well known pictures of the orbits of the Lorenz system. Moreover, the attractive effect of the stable closed orbit seems to persist, allowing both branches of the unstable manifold of the origin to turn around the nearest convective equilibrium, after the homoclinic bifurcation. Following the stable manifold of the symmetrical equilibrium on the opposite side, the unstable manifold of the origin jumps from one leaf of the folded surface to the other one, thus providing a description of the simplest features of the strange behaviour of the orbits in the Lorenz system.

If such a view is correct, we have an example of a strange attractor born after a global bifurcation produced by the folding of a two-dimensional manifold on which the reduced vector field belongs to the class of topological equivalence characterized by system (3). In other words, it seems that in a given n-dimensional system, when the conditions for the previous local codimension-two bifurcation are satisfied, one can confidently guess a chaotic motion in the original system. This is the case for the infinite-dimensional differential equations modelling the original two-dimensional convection problem. There is therefore no need to look for two or three competing physical instabilities, as in the mentioned papers of Arneodo et al.[7], Coullet et al.[8], in order to explain the onset of turbulence. The only convective instability, i.e. the pithfork bifurcation in system (3), together with the closeness of the Hopf bifurcation occurring at the unphysical value $\sigma = -\bar{1}$, suffices to obtain system (3), provided the neighborhood in the parameter space in which the analysis holds is large enough. The other degrees of freedom of the starting equations, and the exponential contraction towards the two-dimensional center manifold may then assure the folding and consequentely the onset of chaos.

To make these statements more precise, we would need an analytical expression for the folding, so that one could describe and justify the three-dimensional bifurcations occurring in the Lorenz system, and particularely those concerning the strange attractor. This will be the subject of our future work.

To conclude, we point out that there should be no surprise when one finds the Lorenz system in different physical contexts: what is behind it is a very ordinary equivalence class, characterized by the structurally stable unfolding of the previous two-zero roots singularity.

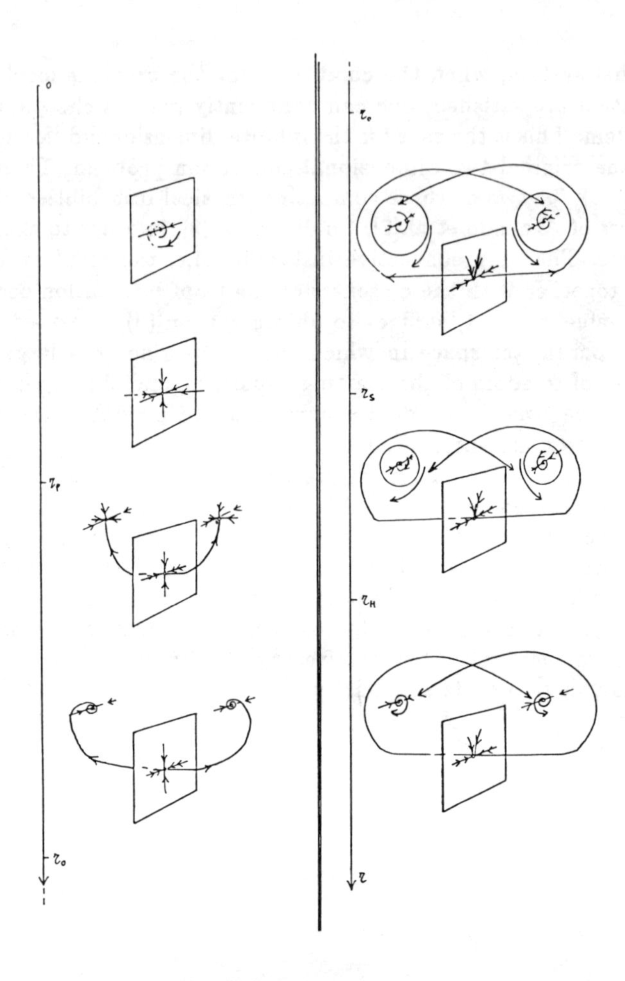

Fig. [1] The modifications leading to the strange attractor observed in the phase portrait of the Lorenz system as the parameter r is varied for σ and b fixed.

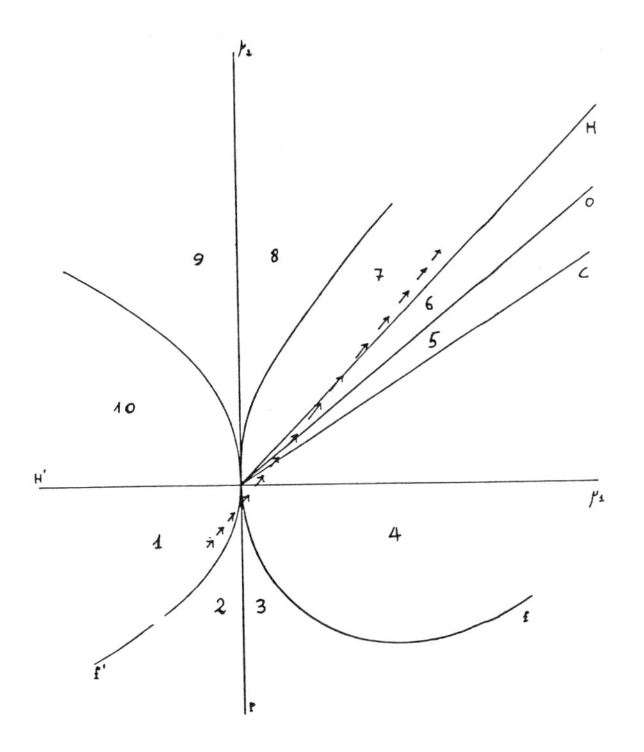

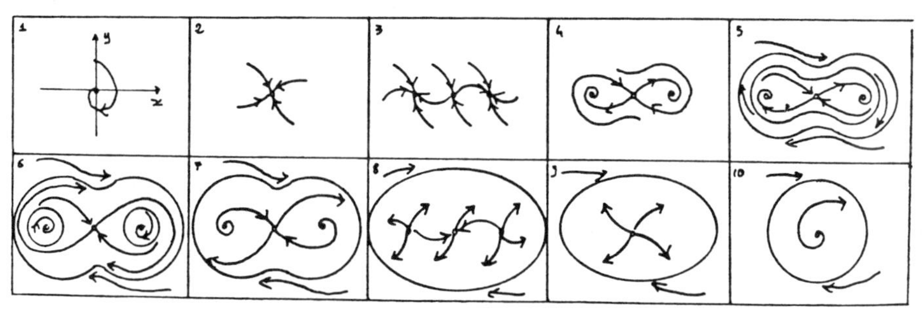

Fig. [2] Phase portraits for system (3), corresponding to the different open sets of structural stability in the plane of the parameters. Also shown is a path in this plane, which reproduces the sequence of the two-dimensional bifurcations actually observed in the Lorenz system.

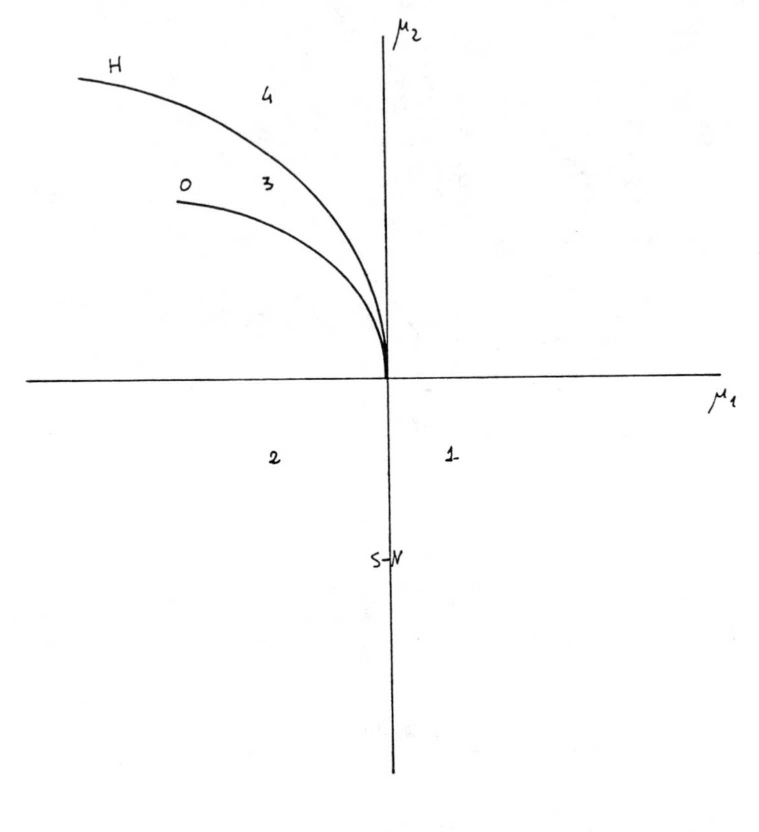

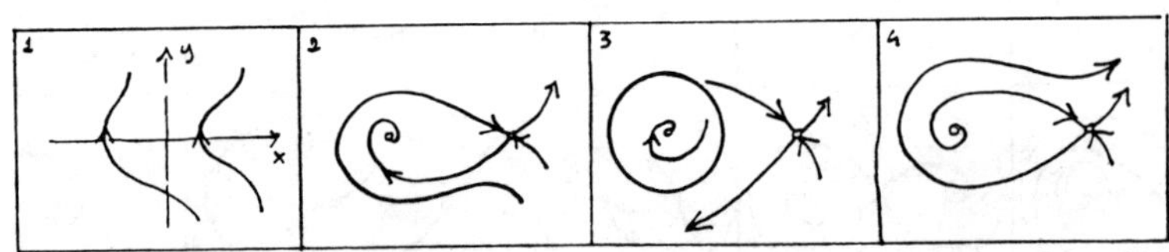

Fig. [3] Phase portraits for system (7) corresponding to the different open sets of structural stability in the plane of the parameters, in the case when the coefficients of the nonlinear terms have the same sign.

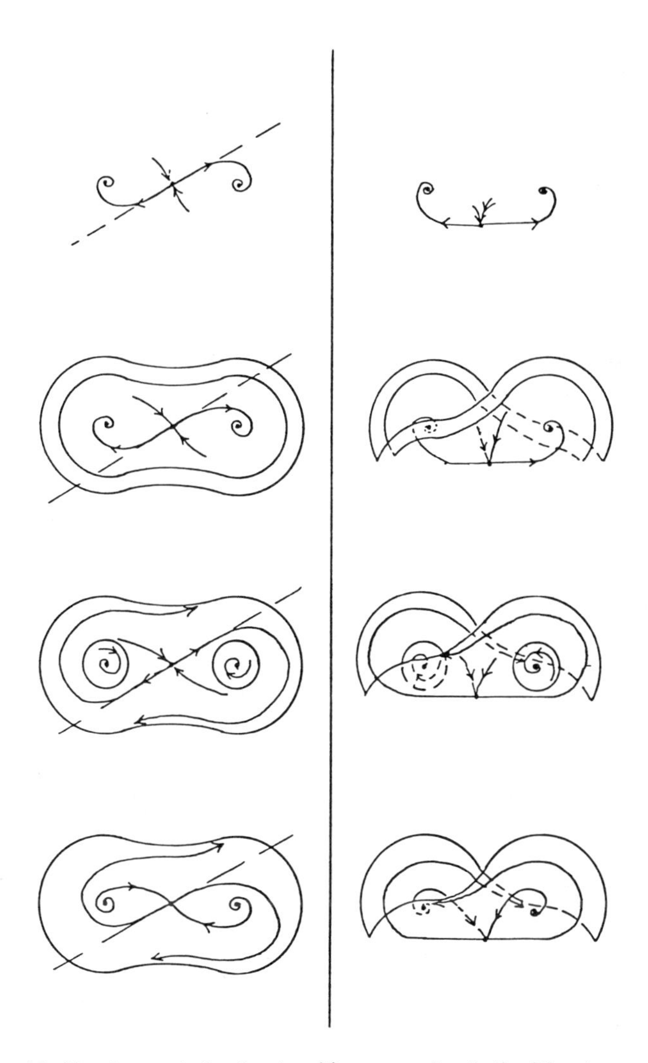

Fig. [4] The phase portraits of system (3), corresponding to the different open sets of structural stability in the upper right quadrant in the plane of the parameters μ_1, μ_2, as they appear when they lie on a plane surface (left column) or on a folded one (right column).

References

[1] E.N.Lorenz, Deterministic Non-Periodic Flow, J.Atm.Sci. 20, (1963) 130-141.

[2] J.Guckenheimer and P.J.Holmes, Nonlinear Oscillations, Dynamical Systems and Bifurcations of Vector Fields (Springer-Verlag, Berlin, 1983).

[3] J.E.Marsden and M.McCracken, The Hopf Bifurcation and Its Applications (Springer-Verlag, Berlin, 1976).

[4] E.N.Lorenz, Noisy Periodicity and Reverse Bifurcation, (1981), preprint.

[5] J.Carr, Applications of Center Manifold Theory (Springer-Verlag, Berlin, 1981).

[6] V.I.Arnold, Chapitres Supplementaires de la Theorie des Equations Differentiales Ordinaires (Editions Mir, Moscow, 1980).

[7] A.Arneodo, J.P.Coullet and E.A.Spiegel, Chaos in a Finite Macroscopic System, Phys.Lett. 92A, (1982) 369-373.

[8] J.P.Coullet and E.A.Spiegel, Amplitude Equations for System with Competing Instabilities, (1983) to appear on SIAM J.Appl.Math.

[9] P.J.Holmes, Center Manifolds, Normal Forms and Bifurcations of Vector Fields with Application to Coupling between Periodic and Steady Motions, Physica 2D (1981) 449-481.

[10] M.Miari, Qualitative Understanding of the Lorenz Equations through a Well Known Second Order Dynamical System, (1984), preprint.

[11] J.L.Kaplan and J.A.Yorke, Preturbulence, a Regime Observed in a Fluid Flow Model of Lorenz, Comm.Math.Phys. 67, (1979) 93-108.

[12] K.A.Robbins, Periodic Solutions and Bifurcation Structure at High R in the Lorenz Model, SIAM J.Appl.Math. 36(3), (1979) 457,472.

INHOMOGENEOUS FRACTALS IN TURBULENCE AND CHAOTIC SYSTEMS

Angelo Vulpiani

Dipartimento di Fisica, Università "La Sapienza" and Gruppo Nazionale di Struttura della Materia del Consiglio Nazionale delle Ricerche, Piazzale A. Moro, 2, I-00185 Roma, Italy

ABSTRACT

The introduction of a multifractal model by a generalization of the standard ß-model allows for obtaining a description of fully developed turbulence in accordance with experimental data; it provides also a clear description of the properties of chaotic systems.

Recently, fractal objects have been introduced in the study of a wide range of phenomena in physics and natural sciences [1]. Usually only homogeneous fractals (i.e. with a global exact dilatational invariace) are taken into account.

An ordinary (homogeneous) fractal is generated by a set of rules which relate its statistical properties over a certain scale of lenghts to those of a larger scale of lenghts. In an inhomogeneous fractal, these rules, at each lenght scale, are not fixed but are instead randomly chosen according to a given probability distribution. While a homogeneous fractal can be thought as an object generated by the fragmentation of a box in a constant number of "son" boxes, an inhomogeneous fractal can be obtained, for instance, changing at random the number of "son" boxes at each step (see fig. 1).

In this report two applications of the inhomogeneous fractals in fully developed turbulence and chaotic systems are shown.

Let us consider the case of fully developed turbulence. It is easy to check that the Navier-Stokes equations:

$$\partial_t \underline{u} + (\underline{u}\cdot\underline{\nabla})\underline{u} = -\frac{1}{\rho}\underline{\nabla} p + \nu \Delta \underline{u} \tag{1}$$

are formally invariant under the scaling transformations:

$$\underline{r} \to \lambda \underline{r} \quad , \quad \underline{u} \to \lambda^h \underline{u} \quad , \quad t \to \lambda^{1-h} t \, , \quad \nu \to \lambda^{h+1} \nu \tag{2}$$

Note that the mean energy dissipation $\bar{\varepsilon} = \nu \overline{(\underline{\nabla} \cdot \underline{u})^2}$ changes as follows under the scaling transformation (2):

$$\bar{\varepsilon} \longrightarrow \lambda^{3h-1} \bar{\varepsilon} \tag{3}$$

The celebrated Kolmogorov's laws [2] can be obtained by imposing the invariance of $\bar{\varepsilon}$ under the scaling transformations (2), i.e. $h = \frac{1}{3}$; this implies that

$$\lim_{\Delta \underline{x} \to 0} \frac{|\Delta \underline{u}|}{|\Delta \underline{x}|^{1/3}} = \lim_{\Delta \underline{x} \to 0} \frac{|\underline{u}(\underline{x}+\Delta \underline{x}) - \underline{u}(\underline{x})|}{|\Delta \underline{x}|^{1/3}} \neq 0 \tag{4}$$

In eq. (4) $\Delta \underline{x} \to 0$ means $\Delta x \sim \eta$ = dissipation lenght, and in the limit of infinite Reynolds number the velocity gradients are singular. Note that in the Kolmogorov theory there is the implicit assumption that the set of singular points (i.e. the points where eq.(4) holds) is filling all the space.

One can think (as Landau [3] first remarked) that the energy dissipation $\varepsilon(\underline{x}) = \frac{1}{2} \nu \sum_{i,j} (\partial_i u_j + \partial_j u_i)^2$ presents strong fluctuations. This happens for example if the set of points where the velocity gradients are singular has a non integer fractal dimension.

An approach to such a problem in terms of homogeneous fractals has been proposed, in the so called ß-model [4]. Roughly speaking in this approach eq.(4) is replaced by

$$\lim_{\Delta \underline{x} \to 0} \frac{|\Delta \underline{u}|}{|\Delta \underline{x}|^{h}} \neq 0 \quad ; \quad h = \frac{D_F - 2}{3} \tag{5}$$

and the set of points where eq.(5) holds has a fractal dimension $D_F < 3$. Practically, in this model, the energy dissipation $\varepsilon(\underline{x})$ is uniformly distributed on a homogeneous fractal object. In the ß-model one obtains for the velocity fluctuation

$$\langle |\Delta \underline{u}(\underline{\ell})|^p \rangle \sim |\underline{\ell}|^{\zeta_p} \quad ; \quad \zeta_p = \left(\frac{D_F - 2}{3} \right) p + (3 - D_F) \tag{6}$$

In fig. (2) we report data on ζ_p for various experimental tests [5]. A linear fit appears reasonable only for $p \lesssim 7$ while for greater values of p there is a tendency for ζ_p to behave in a nonlineare way. Therefore one can conclude that the region where energy dissipates do not constitutes a homogeneous fractal and singularity structures with different h's can be found.

Recently, in order to include more complicated structures a multifractal model has been proposed [6,7].

Let us define S(h) as the set of points for which

$$\lim_{\underline{x} \to \underline{y}} \frac{|\underline{u}(\underline{x}) - u(\underline{y})|}{|\underline{x} - \underline{y}|^h} \neq 0 \tag{7}$$

and let us indicate with d(h) the fractal dimension of S(h). Since the probability to belong to S(h) at scale ℓ is proportional to $\ell^{3-d(h)}$, one has

$$\langle |\Delta u(\underline{\ell})|^p \rangle \sim \int d\mu(h)\, \underline{\ell}^{\,ph+3-d(h)} \sim \underline{\ell}^{\,\zeta_p} \tag{8}$$

and with a saddle point technique one obtains for ζ_p :

$$\zeta_p = \min_h \{ ph + 3 - d(h) \} \tag{9}$$

Eq.(9) physically means that for a given values of p , ζ_p depends on a particular value of h. Hence the kind of instabilities which are needed to set up the sets S(h) are picked out by different moments. Fig. 2 can then be interpreted as the evidence of different instability mechanisms acting on the flow to select the probability distribution of energy transfer and dissipation.

Now we introduce a generalization [7] of the ß-model in order to obtain an inhomogeneous fractal object with different kind of singularities, i.e. different h's, for velocity gradients. Let us consider the scales $l_n = 2^{-n} l_0$ where l_0 is the scale in which energy is injected. If at scale l_n there are N_n active eddies, each eddy $l_n(K)$ generates eddies of size l_{n+1} (k labels the "father" eddy $k=1,\dots,N_n$). Because the rate of energy transfer is constant among $l_n(K)$ and $l_{n+1}(K)$:

$$V_n^3(k)/\ell_n(k) = \beta_{n+1}(k)\, V_{n+1}^3(k)/\ell_{n+1}(k) \tag{10}$$

Here, as in the standard ß-model, $V_n(k)$ is the velocity difference in the active eddy between two points at distance l_n and $ß_{n+1}(k)$ is the percentage of volume occupied by the active eddies of scale l_{n+1} generated by the eddy $l_n(k)$. eq.(10) implies that the velocity difference V_n in an eddy generated by a particular set of fragmentations $ß_1, ß_2, \dots, ß_n$ is

$$V_n \sim \ell_n^{1/3} \left(\prod_{i=1}^{n} \beta_i \right)^{-1/3} \tag{11}$$

From eq.(11) it follows

$$\langle |\Delta \underline{u}(\ell_n)|^p \rangle = \int \prod_{i=1}^{n} d\beta_i \, P(\beta_1 \ldots \beta_n) \beta_i |V_n|^p \tag{12}$$

Since we assume that there are no correlation among different steps of fragmentation, i.e. $P(\beta_1, \ldots, \beta_n) = \prod_{i=1}^{n} \mathcal{P}(\beta_1)$, one obtains for the exponent ζ_p:

$$\zeta_p = \frac{p}{3} - \ln_2 \{ \beta^{(1-p/3)} \} \tag{13}$$

where $\{\cdot\}$ stands for the average over the distribution $\mathcal{P}(\beta)$. Note that, if β is a constant ($\beta = 2^{(D_F - 3)}$ one obtains the results of the standard β-model (i.e. eq.(6)). The knowledge of the probability distribution $\mathcal{P}(\beta)$ is related to the understanding of the nature of the singularities of the Navier-Stokes equations; Fig.(2) shows that the simple form

$$P(\beta) = x\,\delta(\beta - .5) + (1-x)\,\delta(\beta - 1.) \tag{14}$$

leads, for x=.125, to a good fit of the available experimental data, x being the only free parameter. There is no deep reason to choose the form (14) for (β). We have assumed in (14), following a simple phenomenological idea, that an active eddy can generate either velocity sheets (β=.5) or space filling Kolmogorov-like eddies (β=1).

Note that the fractal object generated with the above rules has no more global dilatation invariance properties; nevertheless one can still compute the fractal dimension D_F defined by

$$\langle N_n \rangle \sim \ell_n^{-D_F} \tag{15}$$

where N_n is the number of active eddies at the n-th step of fragmentation. It is easy to show, with a simple calculation (see ref. [7]) that:

$$D_F = 3 + \ln_2 \{\beta\} = 3 - \zeta_0 \tag{16}$$

We remark that the fractal dimension D_F defined in eq.(15) is different from that (D*) often used in the experimental papers and computed by the energy dissipation correlation. In terms of our ζ_p, $D^* = 1 + \zeta_6$. it is easy to see that in general $D_F \geq D^*$; $D_F = D^*$ holds only for the standard β-model (i.e. homogeneous fractal). For instance by the fit given in eq.(14) with x=.125 one obtains D_F=2.91 and D*=2.83:; this is an indirect check of the multifractal nature of fully developed turbulence.

The scenario of the random β-model can be extended to the analysis of dynamical systems [8]. Indeed in the chaotic systems often the motion lies on complicated manifolds of the phase space (the strange attractors) which can have an intricate multifractal structure. In general the fractal dimension D_F cannot fully characterize an attractor, because this is not a homogeneous fractal object. We shall introduce therefore a set of easily computable exponents which generalize the fractal dimension. They are defined in terms of a "local density" n(r) and can be interpreted with the help of a random ß-model.

Let us consider a time series of points $\underline{x}_i=\underline{x}(i\,\Delta t)$ $(i=1,2,\ldots,N \gg 1)$ of the dynamical system

$$\dot{\underline{x}} = \underline{f}(\underline{x}) \quad \text{or} \quad \underline{x}_n = \underline{g}(\underline{x}_{n-1}) \tag{17}$$

where $x,f,g \in \mathbb{R}^d$. The fraction of points contained in a hypersphere of radius r and centered around x_i is:

$$n_i(r) = \frac{1}{N-1} \sum_{j \neq i} \theta(r - |\underline{x}_i - \underline{x}_j|)$$

The moments of such local densities are given by

$$\langle n(r)^q \rangle = \frac{1}{N} \sum_i n_i^q(r) \tag{18}$$

From eq.(18) we define the set of exponents $\Phi(q)$ by the relation

$$\langle n(r)^q \rangle \underset{r \to 0}{\sim} r^{\Phi(q)} \tag{19}$$

In a homogeneous fractal $n(\lambda r)$ has the same statistical properties of $n(r)\lambda^{D_F}$ and this implies:

$$\Phi(q) = D_F \cdot q \tag{20}$$

Generally, attractors do not possess a global invariance under dilatations, and it is only possible to show that (q) is convex in q[9]. Note that $\Phi(1)$ is the exponent ν introduced by Grassberger and Procaccia [10]. It is clear that the $\Phi(q)$ are analogous to the exponents ζ_p. In fig.(3) $\Phi(q)$ is plotted versus q for the Lorenz model and the Henon map. One can see that $\Phi(q)$ is nearly linear at $|q| \lesssim 1$ but deviations from the line $D_F\, q$ appear at large values of q.

It is possible to show that the exponents $\Phi(q)$ are related to the dynamical properties of the systems (17); see for details ref. [8]. the basic idea is that, due to the nonlinear dynamics, the fragmentation of a hypercube of size l_{n-1} in the phase space yields $2^d ß_n$ hypercubes of size $l_n=l_{n-1}/2$ where $ß_n$ is a parameter given by the dynamics and function of the position. Imposing the conservation of the volume (i.e. the number of points) at each step one has:

(21) $$\ell_{n-1}^{d}\, \rho_{n-1}(\alpha) = 2^{d}\, \beta_n(\alpha)\, \ell_n^{d}\, \rho_n(\alpha)$$

where $\rho_n(\alpha)$ is the density in a hypercube of size l_n obtained by a "fragmentation history" α. Note that eq.(21) is the analogue of eq.(10) for turbulent flows. It follows from eq.(21) that

(22) $$\rho_n(\alpha) \sim \left(\prod_{i=1}^{n} \beta_i(\alpha) \right)^{-1}$$

and, recalling that $\langle n(l_n)^q \rangle \sim \langle (l_n^d \rho_n)^q \rangle$, one obtains

(23) $$\Phi(q) = d \cdot q - \ln_2 \{ \beta^{-q} \}$$

Moreover it is easy to show (see ref.[8]) that:

(24) $$D_F = - \Phi(-1)$$

The fractal dimension computed numerically by eq.(24) is in good agreement with the results obtained by the box-counting methods [11].

This report has been based on ref.s[7,8]; I thank the cowokers of these papers for the continuous exchange of ideas about turbulence and chaotic systems. A particular thank goes to G. Paladin with whom all parts of this report have been discussed.

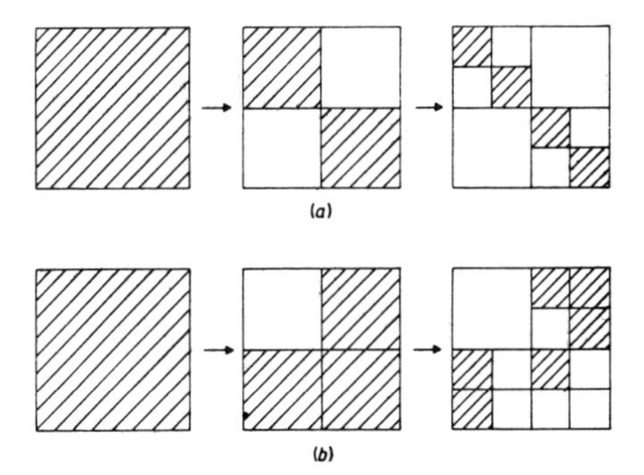

Fig.1 Schematic view of an homogeneous fractal (a) compared with an inhomogeneous one (b). The dashed areas are the zones active during the fragmentation process.

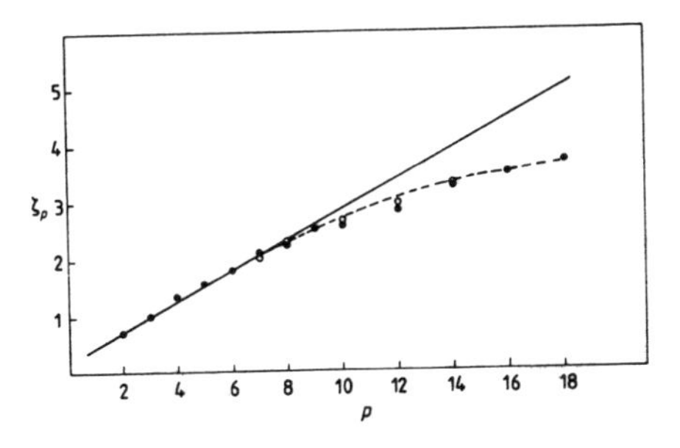

Fig.2 ζ_p versus p. Dots and circles represent experimental data of ref. [5]. Solid line is the ß-model with D_F=2.83. Dashed line refers to eqs. (13) and (14) with x=.125.

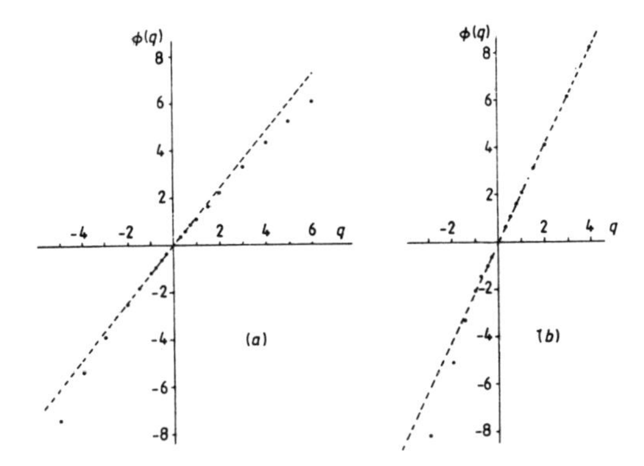

Fig.3 Variation of $\Phi(q)$ against q. Broken lines correponds to the line $\Phi(q)=qD_F$. Dots are $\Phi(q)$ computed with N=10^4 for (a) the Henon map (a=1.2, b=.3) and (b) the Lorenz model, (r=28).

REFERENCES

(1) Mandelbrot B.,
The fractal geometry of nature, Freeman and Co., S. Francisco (1982).
(2) Kolmogorov A.N.,
C.R. Acad.Sci. USSR 30, 301 (1941).
(3) Landau L.D. and Lifchitz E.M.,
Mecanique des Fluides, Moscow M.I.R. ed. (1971).
(4) Frisch U., Sulem P. and Nelkin M.,
J. Fluid Mech. 87, 719 (1978).
(5) Anselmet F., Cagne Y., Hopfinger E.J. and Antonia R.A.,
preprint Institute de Mecanique de Grenoble (1983).
(6) Frisch U. and Parisi G.,
Varenna Summer School LXXXVIII (1983).
(7) Benzi R., Paladin G., Parisi G. and Vulpiani A.,
J. Phys. A (Math.Gen.) 17, 3521 (1984).
(8) Paladin G. and Vulpiani A.,
Lett. Nuovo Cimento 41, 82 (1984).
(9) Feller W.,
An introduction to probability theory and its applications, vol. 2 (Wiley, New York) (1971).
(10) Grassberger P. and Procaccia I.,
Phys. Rev; Lett. 50, 347 (1983).
(11) Russel D.A., Hanson J.D. and Ott E.,
Phys. Rev. Lett. 45, 1176 (1980).

CHAOTIC PATTERN COMPETITION IN A HYDRODYNAMIC SYSTEM

S. Ciliberto (+), J.P. Gollub (++)
(+) Istituto Nazionale di Ottica, Largo E. Fermi 6 - 50125 Firenze (Italy)
(++) Haverford College, Haverford PA 19041, and Department of Physics, University of Pennsylvania, Philadelpia, PA 19104 - U.S.A.

Abstract

Competition between two nearly degenerated spatial modes of different symmetry produces periodic and chaotic fluctuations on the free surface of a circular fluid layer that is forced vertically. The onset of chaos is studied in details. The measurements show that, in the chaotic regime, the attractor has a low (and fractional) dimension and at least one Lyapunov exponent is positive. We also present a four variables phenomenological model whose results are in good agreement with the experimental ones.

Introduction

Some experimental studies of circular Couette flow (1), Rayleigh Benard Convection (2), geostrophic turbulence (3) have demonstrated that the chaotic behavior observed in these hydrodynamic systems, can be described by few dimensional strange attractors(4).

Unfortunately, the physical origin of the chaotic states is not well understood and mathematical models that incorporate the correct dynamics to predict the behavior as a function of external parameters are not generally available.

In a recent paper (5) we described an experiment on forced surface waves in which chaotic behavior is clearly produced by the competition of two spatial modes. We suggested also a phenomenological model that explains many of the experimental results. In this report we discuss the recent progress that we have made in the characterization of the chaotic states. The experimental techniques and the model will be treated in greater details in separate publications (6,7).

Experimental results

The system of interest is a cylindrical fluid layer in a container that is subjected to a small vertical oscillation. It is well known (7) that if the driving amplitude exceeds a critical value $A_c(f_o)$ which is a function of frequency, the free surface develops a pattern of standing waves. The surface deformation $S(r, \theta, t)$ can then be written as a superposition of normal modes.

$$S(r,\theta,t)= \sum_{\ell,m} a_{\ell,m}(t)\ j_\ell\ (k_{\ell,m} r)\ \cos \ell\,\theta \quad (1)$$

where J_ℓ are Bessel functions of order ℓ and the allowed wave numbers $k_{\ell,m}$ are determined by the boundary condition that the derivative $J'_\ell\ (k_{\ell,m}R)=0$, where R is the radius of the cylinder. The modes may be labeled by the indices ℓ (giving the number of angular maxima) and m (related to the number of nodal circles). The mode amplitude $a_{\ell,m}(t)$ develops an instability when the corresponding eigenfrequency (given by the dispersion law for capillary-gravity waves) is approximately in resonance with half the driving frequency f_o and A exceeds A_c. This parametric instability leads to standing waves in which the mode amplitude oscillates at $1/2\ f_o$. To take into account the possibility of a further slow modulation of the mode amplitudes, which in fact occurs due to mode competition, we write each amplitude in terms of fast oscillations at $1/2\ f_o$ and slow envelopes $C_\ell\ (t)$ and $B_\ell\ (t)$

$$a_\ell\ (t)=C_\ell(t)\ \cos(\pi\ f_o t)\ + B_\ell(t)\ \sin(\pi f_o t) \quad (2)$$

We omit the second subscript because in practice only a single value of m is significant for a given value of ℓ.

The behavior of the system as a function of A and f_o is shown in figure 1. Below the parabolic stability boundaries, the surface is essentially flat. Above the stability boundaries, the fluid surface oscillates at half the driving frequency in a single stable mode (Fig. 1 of Ref. 5,6). The shaded areas are regions of mode competition, in which the surface can be described as a superposition of the (4,3) and (7,2) modes with amplitudes having a slowly varying envelope in addition to the fast oscillation at $1/2f_o$.

Our experimental apparatus, described in (5) and (6), allows us to study a fixed linear combination of the slow coefficients $C_\ell\ (t)$ and $B_\ell(t)$, which we denote by $a^o_\ell\ (t)$. This functions, with ℓ = 7 and 4, are shown in Fig. 2. The slow oscillation is periodic in this case with a period of 15 sec.. That is to order of magnitude larger than that of the fast oscillation at $1/2f_o$. We find that $a^o_7\ (t)$ leads $a^o_4\ (t)$ by about 90°. This phase relationship is significant; it implies that energy is being transferred back and forth between the two modes. This result is the basic quantitative evidence for our assertion that the slow oscillations are caused by competition between two spatial modes.

The dynamic of the slow oscillation was explored by varying A and f separately inside of the interaction region. In figure 3 time series and corresponding power spectra are shown for three different driving amplitudes but fixed driving frequency of 16.05 Hz. At A = 121 μm the slow oscillation is periodic. As the driving amplitude is increased the system undergoes a first "noisy" subharminic bifurcation (A = 139 μm) that is clearly seen at A = 149 μm. Beyond 180 μm broad power spectra are obtained as shown in Fig. 3f.

We characterized the chaotic behavior more quantitatively (6) computing from the experimental data the correlation dimension ν of the attractor (9) and a lower bound K_2 for the Kolmogorov entropy K (10). When the oscillation is periodic, A = 121 μm, we find ν = 1.0 + 0.04 and $K_2 = (0.0 \pm 0.01)$ sec^{-1}. On the other hand when the slow oscillation is chaotic (A = 190 μm) ν = 2.22+0.04 and $K_2 = (0.1 \pm 0.01)$ sec^{-1}. The measurements clearly demostrated that the attractor has a low (and fractional) dimension and that there is at least one positive Lyapunov exponent.

Phenomenological model

We have constructed a relatively simple phenomenological model that has a reasonable hydrodynamic basis and accounts for most of our observations, including the basic structure of the phase diagram. We begin with the fact that in a linearized inviscid approximation, each mode amplitude $a_\ell(t)$ follows a Mathieu equation (8). We take the point of view that one can approximately account for the effects of damping (due to all sources, including bulk viscosity and wall effects) by introducing a first order term $\gamma_\ell \dot{a}_\ell$. Furthermore, it is necessary to have a nonlinear term to limit the growth of the mode to finite amplitude in the steady state. The lowest order nonlinear term is cubic in the mode amplitude, so we have the following equation for the time variation of each mode:

$$\ddot{a}_\ell + \gamma_\ell \dot{a}_\ell + (\omega_\ell^2 - \psi_\ell A \cos \omega_0 t)\, a_\ell = \zeta_\ell a_\ell^3 \qquad (3)$$

where ω_ℓ is the eigenfrequency, ψ_ℓ is a gain coefficient, and $\omega_0 = 2\pi f_o$. The coefficients γ_ℓ and ψ^0_ℓ can be adjusted to fit the stability curves and ζ_ℓ are chosen to obtain the correct saturation amplitudes, all in a region where a single mode is present.

We find that this equation is sufficient to fit the (approximately parabolic) stability curves in Fig. 1, and to describe quantitatively the variation of the steady state mode amplitude with A above threshold.

Next, we consider the phenomenon of mode competition. There are various ways to introduce coupling phenomenologically. We assume that the coefficient of the driving term for one mode depends on the amplitude of the other mode. Therefore, to describe the interaction of the $\ell = 7$ and

$\ell = 4$ modes, we set

$$\Psi_4 = \overline{\Psi}_4 + \beta_{47}\, a_7^2 \quad \text{and} \quad \Psi_7 = \overline{\Psi}_7 + \beta_{74}\, a_7^2 \qquad (4)$$

where the coupling coefficient β_{74} is negative while β_{47} is positive. The origin of the sign difference is the observed phase difference between the two modes during the oscillation (see Fig. 2). To solve the system of the two coupled Mathieu equations we express the mode amplitude with Eq. 2. We numerically integrate the resulting four dimensional system for the slow variables $C_\ell(t)$ and $B_\ell(t)$ and find that regenerative oscillations (both periodic and chaotic) are in fact produced near the intersection of the stability boundaries for the two modes. We adjust the two mode-coupling coefficients to obtain an oscillatory domain similar in size to that found in the experiments (Fig. 1). The phase diagram produced by this set of model equations is shown in Fig. 4.

In order to compare the behavior near the onset of chaos with that observed experimentally, we present (Fig. 5) time series of the slow component B and corresponding power spectra for three different values of A but fixed f_o = 16.11 Hz. This figure may be compared with the experimental data in Fig. 3. The basic period of oscillation is different by an unexplained factor of two. However, in both cases we find the following features: a single subharmonic bifurcation of the slowly varying mode amplitude for comparable A and f_o ; an increase in the background noise level at, or near, this bifurcation; and a loss of all sharp spectral structures at higher A without further bifurcations. Thus, the onset of chaos seems to be quite similar in the data and in the model.

We also measured the correlation dimension ν for the chaotic states of the model with the method proposed in (8). We find $\nu = 2.41 \pm 0.04$ at A = 175 μm, and the same result at A = 155 μm. Thus, the strange attractor produced by the model has about the same dimension as that found in the experiment.

The Lyapunov exponents of the model equations were also computed with the method proposed in 11. The resulting Kolmogorov entropy K (the sum of the positive Lyapunov exponents) is 0.33 sec^{-1} at A = 175 μm. The ratio K/f of the Kolmogorov entropy to the frequency f of slow oscillation is approximately 1.1 for both the model and the experiment.

Conclusion

We have shown that the competition between two spatial modes with different symmetry produces periodic and chaotic fluctuations on the surface waves in a vertically oscillating fluid layer.

Many of the experimental results (including the overall structure of the parameter space (Fig. 4), the route to chaos, (Fig. 5), the dimension of the resulting strange attractor, and the Kolmogorov entropry) are fairly well described by a simple phenomenological model. However, the model is not in complete agreement with the data. The period of oscillation is, for example, off by a factor of two and the shapes of the stability boundaries are different. Besides, it lacks of theoretical justifications. So it would be desirable to investigate the relationship between this low dimensional model and the actual hydrodynamic equations, instead of using a phenomenological approach as we have done.

It will be interesting to understand whether pattern competition is a common source of chaos also in other hydrodynamic systems such as Rayleigh Benard convection.

References

(1) A. Brandstater, J. Swift, H.L. Swinney, A. Wolf, J.D. Farmer, E. Jen, J.P. Crutchfield - Phys. Rev. Lett. 51, 1442 (1983).

(2) B. Malraison, P. Atten, P. Bergé, M. Dubois - J. Phys. Lett. 44, 897 (1983).

(3) J. Guckenheimer, G. Buzyna - Phys. Rev. Lett. 51, 1438 (1983).

(4) J.D. Farmer, E. Ott, J.A. Yorke - Physica D 7, 337 (1983).

(5) S. Ciliberto, J.P. Gollub - Phys. Rev. Lett. 52, 922 (1984).

(6) S. Ciliberto, J.P. Gollub - J. Fluid Mech. (to appear).

(7) S. Ciliberto, J.P. Gollub - Nuovo Cimento D (to appear).

(8) T.B. Benjamin, F. Ursell - Proc. Roy. Soc. A225, 505 (1954).

(9) P. Grassberger, I. Procaccia - Physica D 9, 189 (1983).

(10) P. Grassberger, I. Procaccia - Phys. Rev. A28, 2591 (1983).

(11) G. Benettin, L. Galgani, J.M. Strelcyn - Meccanica 15, 9 (1980).

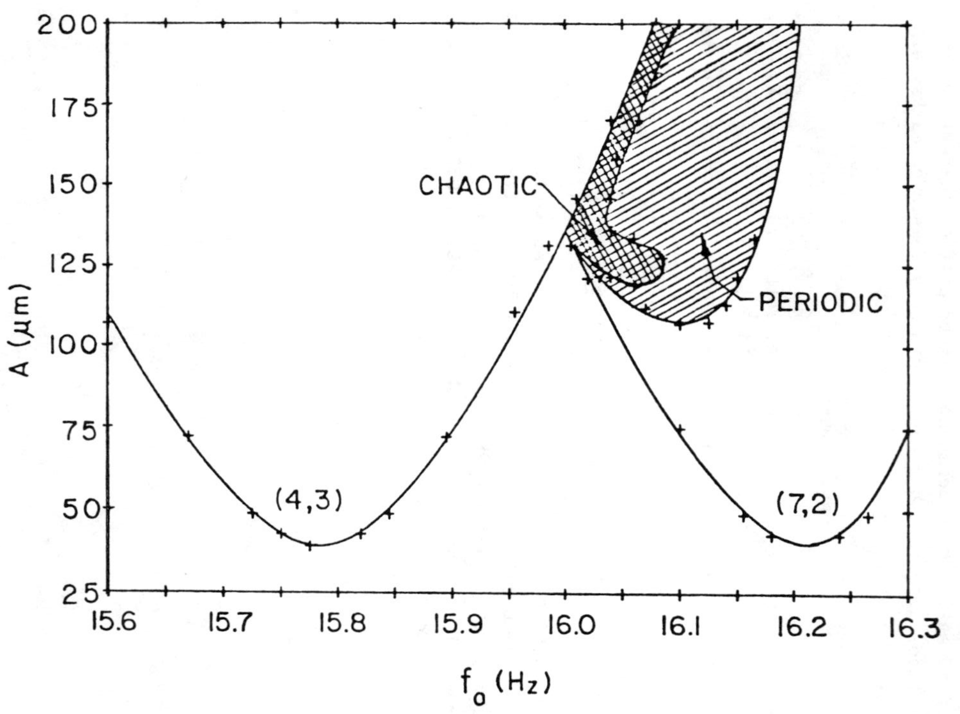

Figure 1. Phase diagram as a function of driving amplitude A and frequency f_o. The crosses are experimentally determined points on the stability boundaries. Stable patterns occur in the regions labeled (4,3) and (7,2). Slow periodic and chaotic oscillations involving competition between these modes occur in the shaded regions.

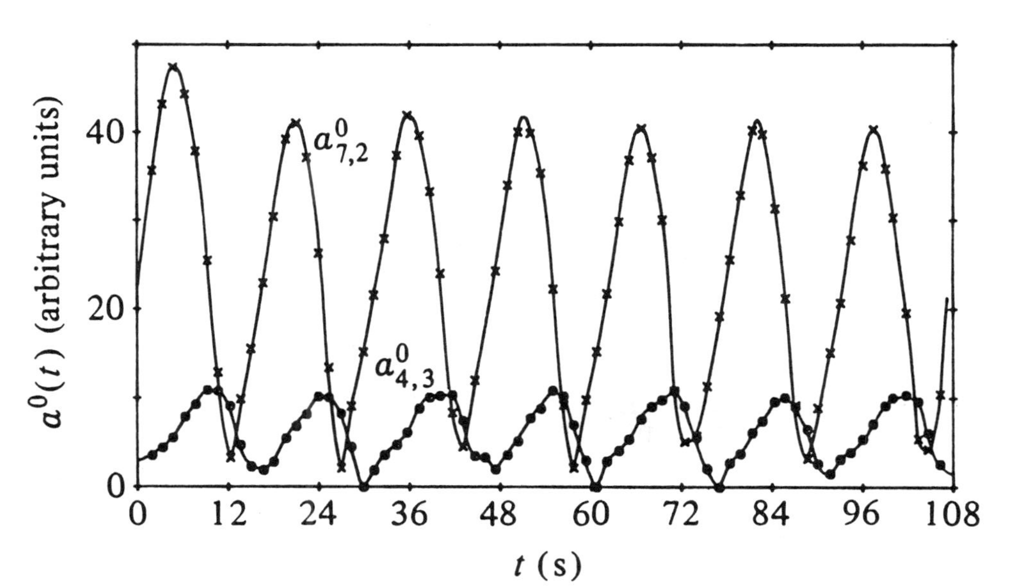

Figure 2. Mode competition. The slowly varying mode amplitudes $a^o_4(t)$ and $a^o_7(t)$ oscillate periodically for f_o = 16.113 Hz. Chaotic oscillations are found at lower driving frequencies.

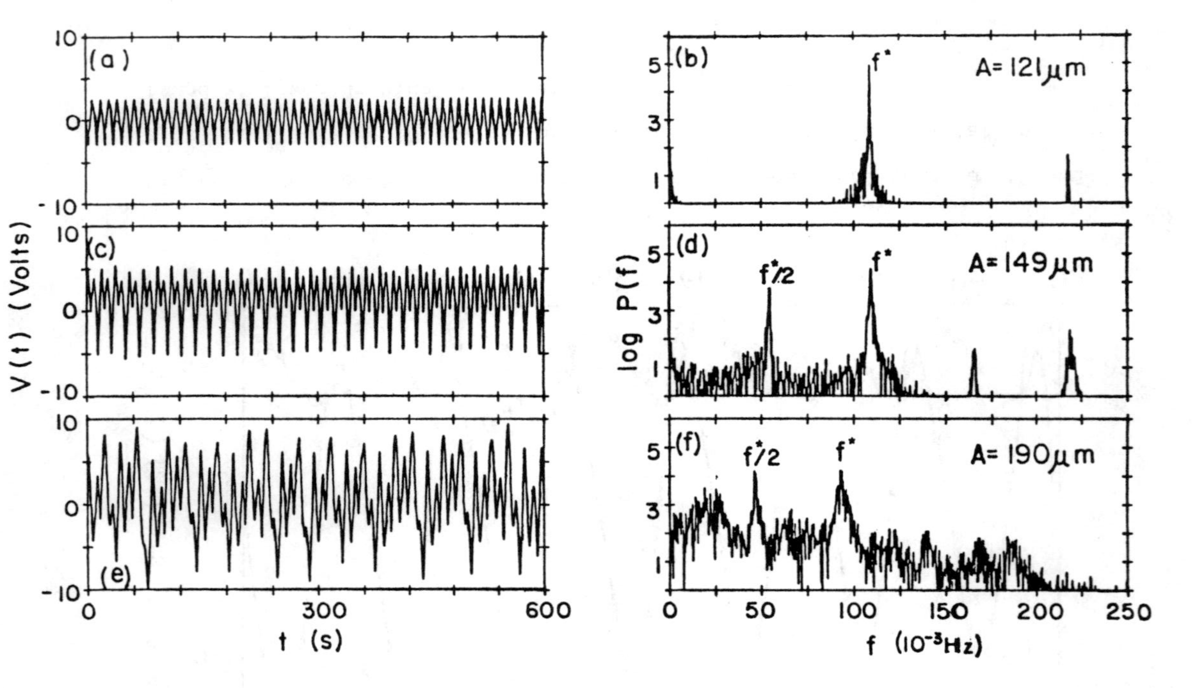

Figure 3. The transition from periodic to chaotic oscillation. Time series and corresponding power spectra of the slow oscillation are shown for f_o = 16.05 Hz and three different driving amplitudes. Broadband noise is associated with the appearance of a subharmonic ½f* of the dominant oscillation.

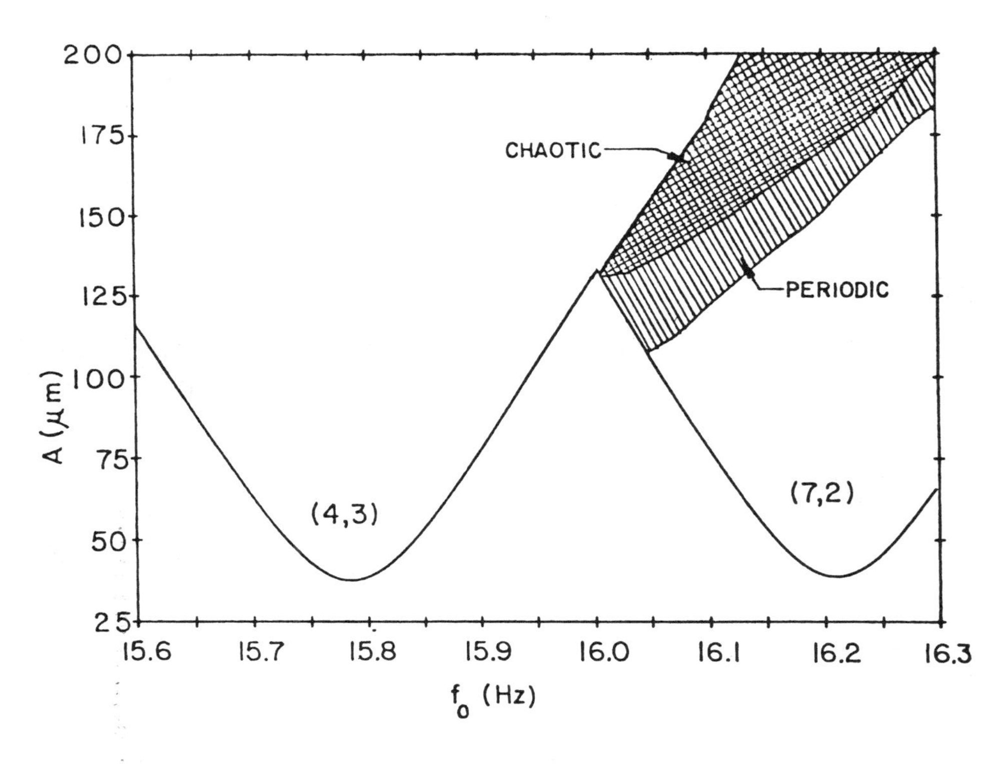

Figure 4. Phase diagram obtained from the phenomenological model by numerical integration. This figure should be compared with figure 1.

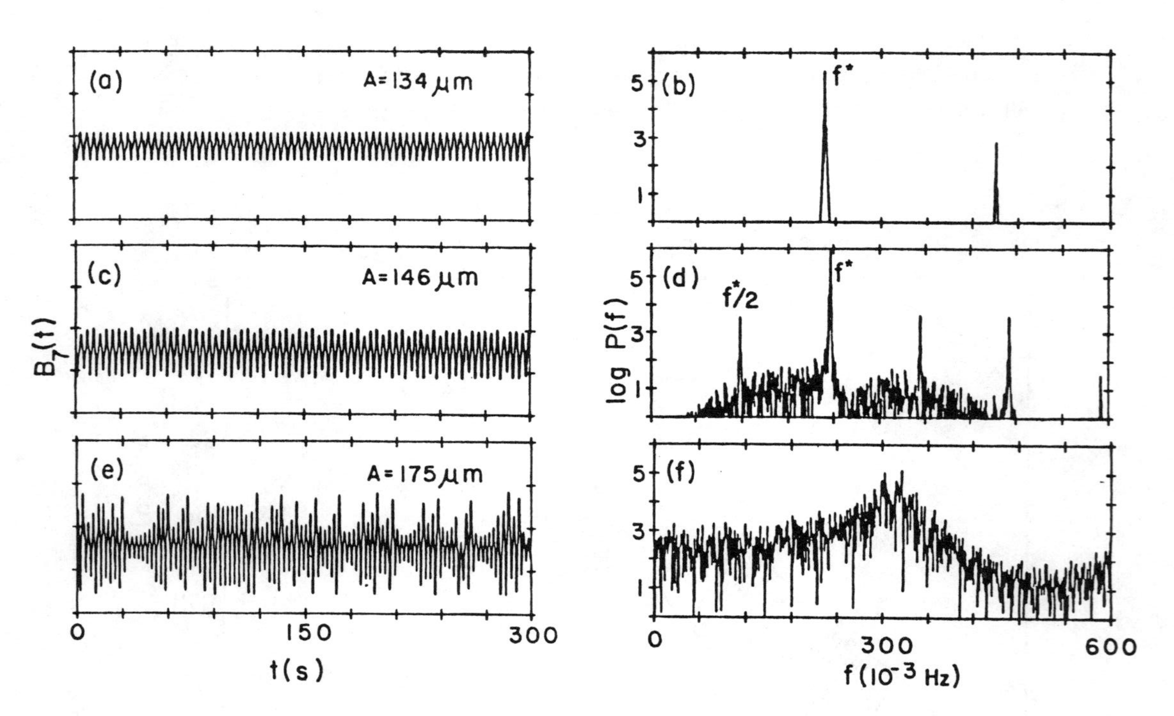

Figure 5. Time series and Fourier spectra obtained from numerical integration of the model at three driving amplitudes. This figure illustrates the transition from periodic to chaos oscillation and should be compared to the experimental data of figure 3. Both experiment and model show a single subharmonic bifurcation with associated broad-band noise onset.

EXPERIMENTAL ANALYSIS OF ATTRACTORS DIMENSION FOR VARIOUS FLOWS

J.G. Caputo

Laboratoire d'Electrostatique et de Materiaux Diélectriques, C.N.R.S., avenue des Martyrs,
166 X 38042 Grenoble Cedex (France)

ABSTRACT

We have applied the Grassberger Procaccia method - for determining attractors dimension - to various experimental situations: Rayleigh-Benard convection, grid turbulence and Electrohydrodynamical convection. A systematic study makes a clear cut difference between a Low-dimensional attractor ($\nu \sim 3$ or 5) and a very high-dimensional system. Several effects are pointed out, which may be useful in future studies.

Introduction

For a fluid experimentalist, the important question when analysing aperiodic data is to see whether the dynamics is deterministic or stochastic. To see this, one can calculate the dimension of the attractor associated to the signal, by using an algorithm recently proposed by Grassberger and Procaccia. [1]

In this paper, I shall present results of the application of this method to different flows and point out several difficulties.

1. Method for calculating the dimension for a flow.

From a time-series on a single scalar observable $X(j\cdot\Delta t)$ $(j = 1, N)$ of the experiment we reconstructed an n-dimensional phase space by the now-classical method of timedelays $(\underline{X}(t) = (X(t - p\cdot\Delta t), \ldots, X(t - p(n)\cdot\Delta t))$. For attractors of Hausdorff dimension d_H, this representation is surely an embedding for $n > 2d_H + 1$ [2]. To estimate the dimension of the attractors we adapted the method in [1]. We computed:

$$(1) \qquad C_i(r) = \frac{1}{N} = \# \left\{ \text{points } \underline{X}(j\cdot\Delta t) \text{ such that } \|\underline{X}_i - \underline{X}(j\cdot\Delta t)\| \leq r \right\}$$

and to smooth out $C_i(r)$ and get better statistics:

$$(2) \qquad C'(r) = \frac{1}{m} \sum_{i=1}^{m} C_{i(r)}$$

Assuming a nearly homogeneous fractal structure - which was true for every attractor studied here - we shòuld have:

$$(3) \qquad C_i(r) \underset{r \to 0}{\sim} r^{\nu}$$

ν independent of i, being it's dimension. As all norms are equivalent in R^n, the limit in (3) does not depend on the choice of the norm and we have verified this numerically. So, for practical reason, we have only used the maximum norm. The studies below were all made with $150\ 10^3 \leq N \leq 250\ 10^3$ and $300 \leq m \leq 1000$ except for the Rayleigh-Bénard case where $N = 10^4$. Experimental signals were digitalized on 12 bits and numerical signals on 15 bits. The instrumental noise has no effect on the estimates below [3]. The sampling rate was between 30 and 40 points per pseudo-period, except for the grid turbulence case (100 points per period)..

2. Rayleygh Bénard Convection

The experiment was made by Bergé and Dubois in a parallelepidic box with aspect ratios $\Gamma x = 2$, $\Gamma y = 1.2$ and silicon oil ($Pr \simeq 40$). The chaotic state analysed followed a period doubling sequence. The semi-local variable is the deflection of a light beam passing through the cell; this enabled for a ratio signal/noise over 10^5 in power. A first study [4] had indicated the presence of a strange attractor, but the parameters n, $p \cdot \Delta t$ had been chosen such that $n \leq 12$ and $(n - 1)\ p \cdot \Delta t = cst$. This was made to curb the effects of decorrelation that occur when n becomes too large. Here the systematic variation of n from 10 to 30 and of p from 1 to 20 clearly shows an independence of the dimension. (Fig. 1).

These results are immediately comparable with those obtained for the Lorenz system and confirm the previous diagnosis of a strange attractor.

3. Grid Turbulence

The experimental situation analysed by Gagne [3] consists of a turbulent flow behind a grid. This system is known to possess a very high number of degrees of freedom. The local variable analysed is a component of

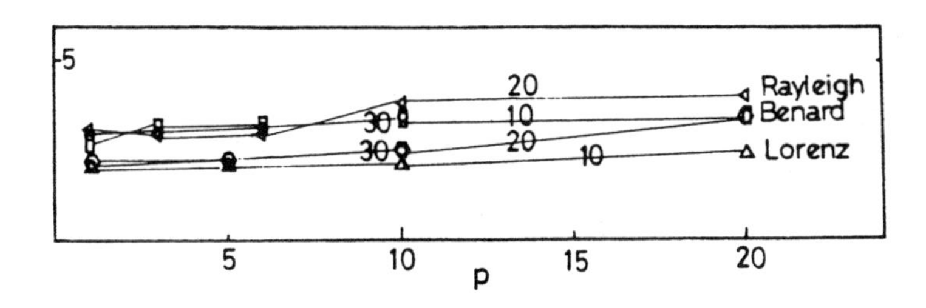

Fig. 1 : ν as a function of p for different dimensions n:
- Rayleigh Bénard convection regime
- Z(t) variable of the Lorenz system

The slight deviations when n or p get larger arise because of the reduction of the scaling region due to decorrelation; but, looking at the local slopes - as advocated by Ciliberto - reveals a strict saturation.

the velocity at a given point behind the grid; for a fixed embedding dimension n, the exponent ν tends towards a limit value ν_s as the time-delay increases (Fig. 2). For this very high-dimensional system, the autocorrelation drops very quickly and the different coordinates become inde-

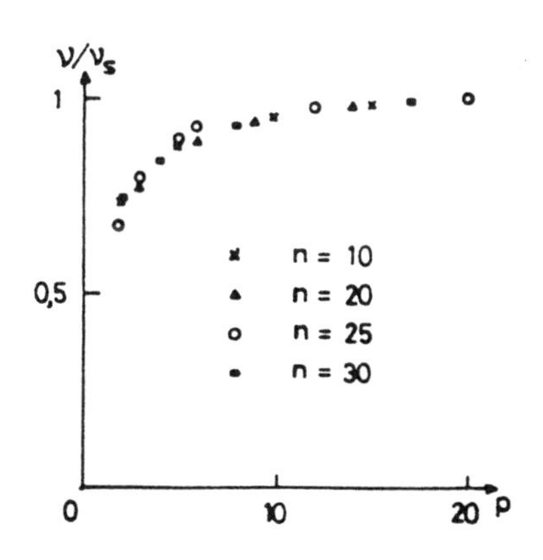

Fig. 2 : The apparent dimension rate ν/ν_s as a function of the time delay P for various n.

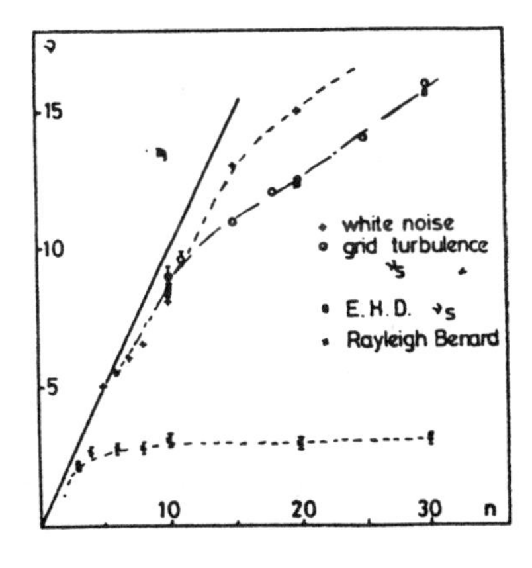

Fig. 3 : Variation of ν with the embedding dimension n for different experimental signals.

pendent one with another for a small value of p. There is no "geometric correlation between the vectors and the dimension of the cloud of points keep growing with n (Fig.3).
White noise tends to be more space filling for a given laps of time than coloured noise. This may explain the difference observed for the dimension between white noise and the grid turbulence signal.

Another interesting feature is the existence of "tails" of slope ~ 1 for the small values of r in the curves C(r) (Fig. 4). At very high dimensions, and time delays the point $\underline{X}(i \cdot \Delta t)$ has as nearest neighbors the points coming immediately before and after him. The "dimension" seen at that scale is one. This effect increases with the sampling frequency.

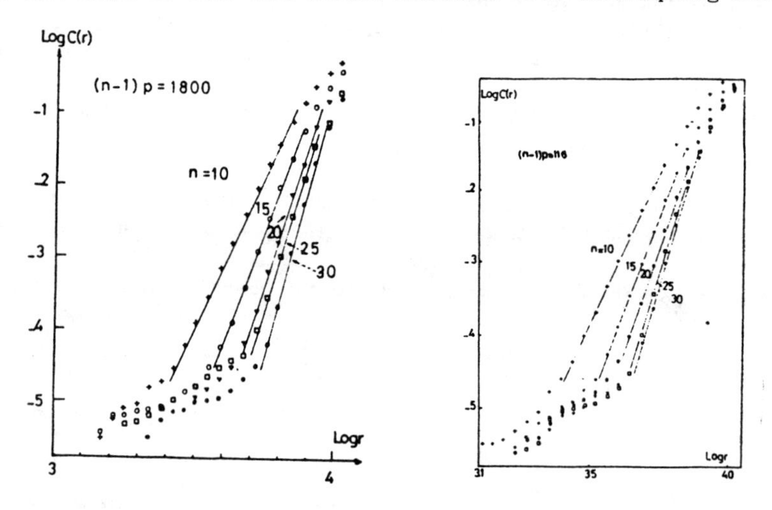

Fig. 4 : Different curves C(r) for grid turbulence.

4. Electrohydrodynamical Convection (E.H.D.)

Consider a layer of insulating liquid submitted to an ion injection due to a voltage difference applied to the liquid.
It can be shown that the charge distribution with the fluid at rest is potentially unstable. An instability develops and the driving parameter T is proportional to the voltage. Due to the symmetry the convection cells are here hexagons. The global variable analysed is the total electric current crossing the cell; it is related to the integral of the velocity field.

For a smal cylindrical aspect-ratio the transition to turbulence is typical of the Ruelle-Takens picture [5]. Furthermore, the power-spectra present the exponential decrease wich seems to be caracteristic of deterministic chaos [6].

A study made in the same conditions as for the thermal convection experiment, tended to indicate an attractor of dimension around 5. However, this result appears clearly erroneous when one varies systematically the embedding dimension n and delay p (Fig. 5).

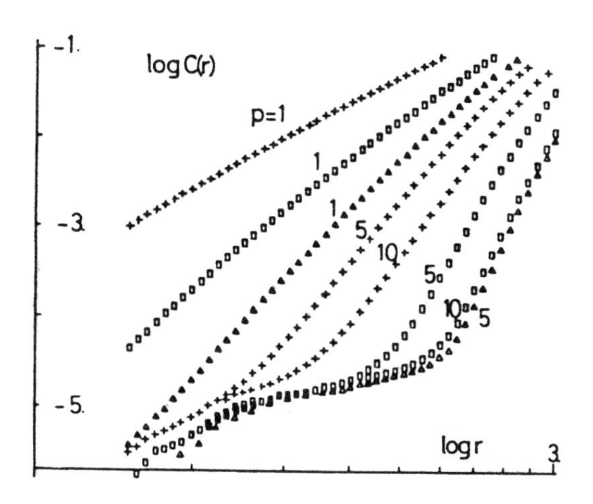

Fig. 5 : Curves C(r) versus r in log-log plots for E. H.D. for n=10 (+),20 (□),30 (Δ) for different delays p.

Surprisingly, this system, for the dimensions analysed, behaves much like the grid-turbulence signal (Fig. 3). We observed the same saturation of ν with p. (Fig. 2). It is not yet well understood why no low-dimensional attractor is foud even for an aspect-ratio of 1 and a small value of the instability parameter.

5. Concluding remarks

This study shows that only a systematic variation of the parameters n,p, Δt can allow for an identification of low-dimensional attractor.

This has permitted a clear confirmation of the presence of a low-dimensional attractor ($\nu \sim 3$) for the Rayleigh-Benard experiment. For the

E.H.D. as for the grid-turbulence the dynamics is supported by a very large number of degrees of freedom ($\nu > 16$).

When trying to characterise such systems, one has to vary the sampling frequency in order to avoid the "tails" present in the curves C(r) when embedding the system in very high-dimensional spaces. But, increasing Δt too much results in decorrelation of the vectors $\underline{X}(t)$ for high values of r and this will reduce the scaling region. This effect might lead to an erroneous estimate of the dimension because of a monotonic curvature of the curves C(r).

REFERENCES

1 P. Grassberger and I. Procaccia, Phys. Rev. Lett., 50, p. 346 (1983)

2 F. Takens, "Dynamical Systems and Turbulence, Warwick 1980", Lecture Notes in Math. n. 898, Springer, p. 366 (1980)

3 P. Atten, J.G. Caputo, B. Malraison and Y. Gagne Journal de Mecanique, Special Issue "Bifurcations and Chaotic Behaviour" (1984) in press.

4 B. Malraison, P. Atten, P. Bergè and M. Dubois Comptes Rendus de l'Academie des Sciences, 297 II, (1983) p. 209 and J. Physique Lett., 44, (1983), L897

5 B. Malraison and P. Atten in "Symmetries and Broken Symmetries" Pub. N. Boccara (IDSET, Paris) p. 439 (1981)

6 B. Malraison and P. Atten, Phys. Rev. Lett. (49) p. 273 (1982)

INSTABILITIES AND CHAOS IN LASER SYSTEMS: A REVIEW OF STUDIES ON CO_2 LASERS

F.T.Arecchi[a], N.B.Abraham[b], W.Gadomski[c], G.L.Lippi[d], R.Meucci[e], A.Poggi, G.P.Puccioni[d], N.Ridi, J.R.Tredicce[f].

Istituto Nazionale di Ottica. L.go E. Fermi 6, I50125 Firenze, Italy

ABSTRACT

We discuss the appearance of chaotic instabilities in lasers as examples of low dimensional chaos with a good correspondence between theoretical model and experiments. Use of a homogeneous gain line and single mode operation restricts the dynamics to Lorenz-like equations. We define various classes of laser dynamics and in particular review our experiments on CO_2 lasers either with modulated parameters(loss or gain) or in interaction with another e.m. field detuned from the internal operation frequency .In the two cases chaotic phenomena are observed when the modulation frequencies or the detuning, respectively, are close to the internal resonance frequency.

Introduction

The laser dynamics, arising from the quasi resonant interaction between a single mode e.m. field and an excited medium with a homogeneous gain line, is described by the equations coupling five macroscopic variables, namely, the complex field amplitude E, the complex medium polarization P, and the real population inversion D. At exact resonance we can reduce the description to three real variables(E, P, D) since the phases play no role. Furthermore a simpler description is obtained assuming the mean field approximation and unidirectional single longitudinal mode operation. The single mode operation is obtained by using a inter-mirror distance L such that the frequency separation c/2L between longitudinal modes is larger than the atomic width $\Delta\nu_a$, and rejecting the transverse modes by working near the diffraction limit. Even so, the interplay between $\pm\vec{k}$ components of the standing wave would give rise to higher space harmonics. To further simplify the dynamics we consider a unidirectional field, thus the equations couple the amplitudes of a single $\vec{k}$ wave as follows (1)

$$\dot{E} = -kE + gP \qquad \text{1a}$$
$$\dot{P} = -\gamma_{\perp} P + gED \qquad \text{1b}$$
$$\dot{D} = -\gamma_{\parallel}(D-D_o) - 4gEP \qquad \text{1c}$$

where K, $\gamma_{\perp}$, $\gamma_{\parallel}$ are the relaxation rates of field, polarization and population inversion, respectively, D_o is the population imposed by the excitation mechanism ("pump") in the absence of field, and

$$g^2 = \frac{\omega\mu^2}{\hbar\varepsilon_o V}$$

(V being the cavity volume, ω the transition frequency and μ the matrix element of the dipole operator) is the coupling constant in M.K.S. units.

A laser represented in this form is completely equivalent (2) to a Lorenz system (3). So far we should conclude that every laser must show unstable behaviour for some values of the control parameters, which is in contradiction to all past experience on lasers.

In order to explain this point we need a deeper insight into the physical processes which take place during laser action. Eqs. 1 show that, besides the coupling, the three variables have their own relaxations. If one variable relaxes much faster than the other ones we can take the stationary solution for that variable, which in fact is still slowly varying because of the coupling, and reduce to a smaller number of coupled differential equations (adiabatic elimination of the fast variables (4)).

This is actually the case for a large number of lasers. In fact in many systems polarization and population inversion have relaxation times much shorter than the cavity lifetime ($\gamma_\perp, \gamma_\parallel \gg K$) and both variables can be adiabatically eliminated. With just one variable describing the dynamics, the laser must show necessarily a stable behaviour (fixed point in phase space). This group of lasers has been classified as Class A (5) and contains many common systems (He-Ne, Ar^+, Dye...). In some other cases (Class B) only polarization is fast ($\gamma_\perp \gg \gamma_\parallel, K$) and hence two variables are necessary for the dynamics. In this class we find ruby, Nd and CO_2 lasers which exhibit oscillating behaviour under some conditions, although ringing is always damped.

It is then clear by now that most lasers are not described by the full set of eqs. 1. So that it is not possible to obtain chaotic behaviour from a ring laser so long as the conditions above mentioned are verified. A possible exception is constituted by far-infrared lasers (FIR). For long wavelength molecular transitions the three relaxations constants may be of the same order of magnitude (Class C lasers)(6).

The CO_2 laser

As mentioned above, the CO_2 laser is of class B. The relaxation rates are: $\gamma_\perp \sim 10^8$ sec^{-1}, $\gamma_\parallel \sim 2.5\ 10^3 sec^{-1}$ (collisional(*)), and $K \sim 10^7$ sec^{-1} (typical value for a 3 m long resonator with 20% transmittance at the coupling mirror). We show the reason why the field can not be adiabatically eliminated from eqs. 1. Let us give a look at the stability of the solutions for a CO_2 laser by means of a linear stability analysis.

Performing the adiabatic elimination of polarization (eq.1b) and defining $x=(2g/\sqrt{\gamma_\parallel \gamma_\perp})E$, $z=(g^2/\gamma_\perp K)D$ and $z_o=(g^2/\gamma_\perp K)D_o$ eqs.1 reduce to

(*)Linewidth $\gamma_\perp$ and amount of homogeneous to inhomogeneous broadening can be controlled by the total gas pressure and mixture ($He:N_2:CO_2$) composition. This laser is usually operated in a region where the broadening is essentially homogeneous, resulting in a great advantage for modelling and physical interpretation.

$$\dot{x} = -Kx(1-z) \qquad 2a$$
$$\dot{z} = -\gamma_{\parallel}[z(1+x^2)-z_o] \qquad 2b$$

The secular equation for this system is

$$\lambda^2 + \gamma_{\parallel} z_o \lambda + 2\gamma_{\parallel} K (z_o - 1) = 0$$

with solutions

$$\lambda = -\frac{\gamma_{\parallel} z_o}{2} \pm \sqrt{\frac{(\gamma_{\parallel} z_o)^2}{4} - 2\gamma_{\parallel} K(z_o - 1)} \qquad 3$$

They are always stable because the first term is always larger or equal to the square root, but the actual behaviour depends on the relation between $\gamma_{\parallel}$ and K. It is easily seen that for $\gamma_{\parallel}<2K$ (condition well verified in a CO_2 laser) the square root has imaginary values and is responsible for oscillations in the field. This is the reason why the dynamics of the CO_2 laser must be described by two variables, and the field, although relaxing with a much higher rate, can not be adiabatically eliminated.

The frequency of the relaxation oscillations is approximately (**)

$$\Omega = \sqrt{2\gamma_{\parallel} K(z_o-1)} \qquad 4$$

In order to increase the number of relevant variables of a CO_2 laser from two to three or more, we have recurred to two classes of laboratory expedients:

a) modulation of one characteristic parameter

b) interaction with a nonresonant e.m. field.

The first class is simpler and easier to control experimentally, while the second one gives rise to a broader phenomenology although more difficult for laboratory feasibility. Notice that both classes, besides making it possible the realization of a low dimensional chaos, correspond to devices currently used in the laboratory practice.

Parameter modulation

Defining the intensity $I=x^2$ eqs.2 can be written as

$$\dot{I} = -2KI\ (1-z) \qquad 5a$$
$$\dot{z} = -\gamma_{\parallel}[z(1+I)-z_o] \qquad 5b$$

A new degree of freedom can be introduced by modulating one of the two external parameters occurring in the equations: cavity losses (K) or pump rate (z_o). It is not convenient to consider $\gamma_{\parallel}$ because, even though it can be affected by external means, its experimental control is not very reliable.

We first examine loss modulation and then make a comparison with pump modulation. The third degree of freedom for eqs. 5 is obtained by making the system non autonomous with K time dependent

$$K(t) = K_o(1 + m \cos\omega_m t) \qquad 6$$

where m is the amplitude of the modulation, ω_m its frequency and K_o the unperturbed relaxation rate. If ω_m is far from Ω the system follows the

(**)For a more detailed stability analysis see Ref. 5b.

sinusoidal variation of K, but if ω_m is near to Ω (or to one of its harmonics) nonlinear resonances are excited even for small m. Experimentally we obtain this modulation by an electro-optical modulator in the cavity. The frequency is accurately controlled by a microprocessor driven synthesizer and can be varied in steps as small as 1 mHz without phase change.

Our experiment (7) shows that the first unstable region occurs around 64 KHz, and that new windows appear at higher harmonics, with smaller modulation depths. Various attractors, each with its proper periodicity, can coexist in the same parameter region and the system can enter one or another depending just on the path it has followed. Noise induces jumps away from less stable attractors or between equivalent attractors. As a matter of fact the great accuracy (with no phase jump) of the frequency synthesizer has allowed us to enter some poorly stable and narrow attractors not reacheable in a rougher experiment.

We fix the frequency at 191 KHz and change m from 1% to 20%. By increasing m we obtain a Feigenbaum cascade: f,f/2, f/4, f/8, chaos, f/3,...allowing a rough evaluation of Feigenbaum's constant δ_F =3.7±.2. It is to our knowledge the first time that Feigenbaum's scenario is well reproduced - and not just guessed from some characteristics - in laser systems. This is due to the high experimental reliability and control accuracy obtained.

The laser output intensity, in this region was digitized with a high sampling frequency to retrieve the signal as shown in Fig.1(left) or synchronizing the sampling with the modulation frequency to obtain the projection of the Poincarè Section (only one point per period), Fig.1(right).

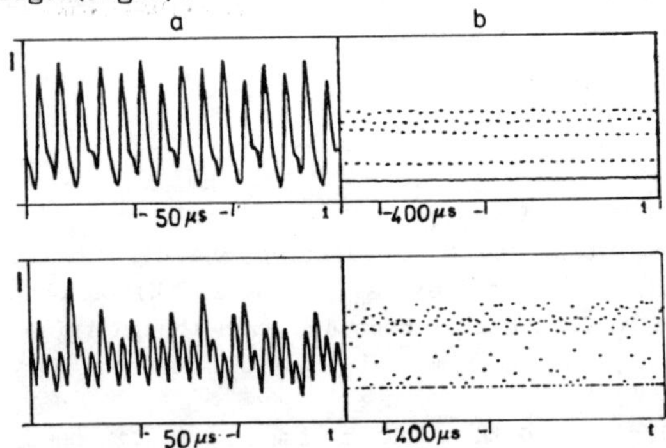

Fig.1 a) Laser intensity vs. time, b)stroboscopic intensity plot for an f/8 subharmonic (above) and chaos (below).

We analyze these sequences reconstructing the attractor by an embedding technique (8,9) allowing the evaluation of its dimension. Following the method of Grassberger and Procaccia (10) we found, as predicted, a dimension near to 1 (0 for the Poincarè Section) for periodic behaviour,Fig.2a-b. When the system enters the chaotic region the fractal dimension jumps to a higher value (between 2 and 3 for the time series and between 1 and 2 for the Poincarè Section) Fig.2e-f, according with the general theory of strange attractors (11) . In Fig.2c-d we see that for the f/8 subharmonic the dimension is near 1.5 (.5 for the Poincarè Section). This result even though not readily understandable because the time signal appears to be

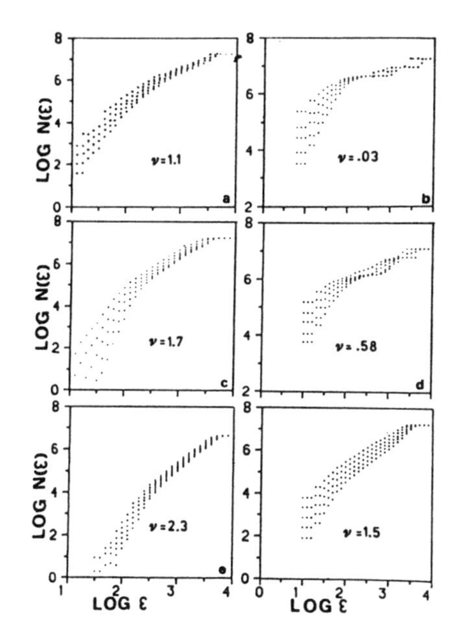

Fig.2 Plots of log $N_n(\epsilon)$ vs. log(ϵ)for different values of the embedding dimension n calculated from the time series (left) and from stroboscopic sections (right) for f/4 (a,b), f/8 (c,d), and chaos (e,f).

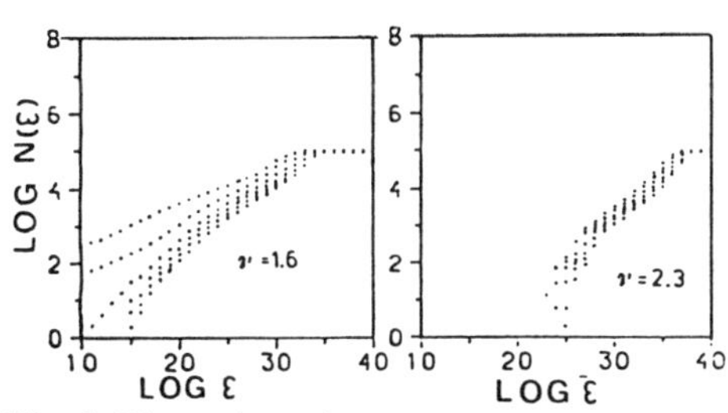

Fig.3 Dimension plots obtained from numerical integration of the model for f/8 (left) and chaos (right).

periodic, nevertheless agrees with the theoretical prediction of the dimension at the accumulation point of the logistic map (12).

In our experimental system noise yields a trajectory wandering over a nonzero range of parameter values, thus "testing" nearby periodic attractors of the subharmonic sequence (13) and for an f/8 subharmonic, which is the limit between periodic and aperiodic behaviour, the dimension is that of the accumulation point.

The high regularity of the dimensions calculated on the stroboscopic plot (Figs.2b and f) suggests the application of a method proposed (10) to give a lower bound (K_2)of the Kolmogorov entropy. We have found that, while K_2=0 for f/4, K_2 ~ 35kHz for the chaotic attractor.

Numerically integrating the model and processing the data obtained in the same manner we obtained results in agreement with the experimental ones. Direct comparison of Fig.3 with Fig.2c and e shows that the agreement is accurate.

Let us consider now pump modulation (14)

$$z_o(t)=A_o(1+\ell \cos \omega_\ell t) \quad 7$$

From eqs. 5 we see that pump or loss modulation forces in a similar way the same nonlinear oscillators, thus time behaviour must be similar in both cases(15). We have shown however (14) that the effective forcing amplitude for the two cases is given by

$$\frac{m}{\ell} \sim A_o \left[\frac{\gamma_{/\!/}}{2K_o(A_o-1)}\right]^{1/2} \quad 8$$

and taking A_o ~ 2 we obtain $m/\ell \sim 10^{-2}$ .This means that the modulation depth must be scaled correspondingly and hence for pump we must use a modulation depth much larger than in the other case: 2.9% of loss modulation is equivalent to 53.6% pump modulation.

Furthermore for large modulation some of the assumptions made in deriving the fundamental equations break down.

Interaction with a nonresonant e.m. field

We may increase the number of degrees of freedom by letting an extra e.m. field interact with the single mode laser. We discuss two different ways to do so: either by injecting a detuned external field or by building a ring laser in which both directions are allowed.

Let us start with the first system. Here it is necessary to take into account the detuning $\omega_{ei} = K\theta$ between the external and the internal field, and, for completeness, also that between the cavity mode and the molecular transition frequency $\omega_{im} = \gamma_{\perp}\delta$. The equations become (5)

$$\dot{x} = (z-1)x + (\theta - \delta z)y + A \qquad 9a$$
$$\dot{y} = -(\theta - \delta z)x + (z-1)y \qquad 9b$$
$$(K/\gamma_{\parallel})\dot{z} = (z-z_0) - z(x^2+y^2)(1+\delta^2)^{-1} \qquad 9c$$

where x and y are the real and imaginary part of the complex field amplitude and A is the real amplitude of the injecting field. Although in the equations the polarization has already been adiabatically eliminated, the presence of two independent field components yields three relevant dynamical variables. Here below we discuss the numerical solutions of this model, the experiment being under preparation.

The system shows bistability in the output intensity with respect to the injecting intensity, with an hysteresis amount which depends on both detunings (Fig.4). The lower part of the bistable curve is always unstable. The upper part is wholly stable for zero detuning ($\theta - \delta = 0$). For nonzero detuning, it has a stable locked region and an unstable one where the laser oscillates either regularly or irregularly. We have observed two different ways to reach the locked regime, either by decreasing the oscillation frequency (tangent bifurcation), or by decreasing the amplitude of oscillations (Hopf bifurcation), and two different routes to chaos, either by intermittency or by period doubling.

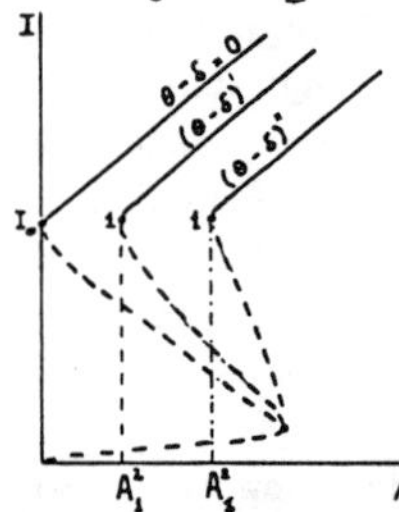

Fig.4 Steady state solutions: output intensity vs. intensity of the injected signal for constant z_0 and different detunings.

In Fig.5 we show regular oscillations in the output intensity: the higher frequency is that predicted by the linear stability analysis, (eq. 4) while the lower one is related to a spiking - with amplitudes which can be also ten or more times higher than the steady state - due to field injection. As a matter of fact we have found that in this region, when the injecting amplitude is too low to lock the system steadily, the laser operates for most part of the time at ω_e(external frequency) but it regularly unlocks going to ω_i (internal frequency). During the oscillation at ω_i the energy of the injecting field enhances the population inversion so that it gives rise, with a delay related to the injecting intensity, to a giant pulse. Increasing the

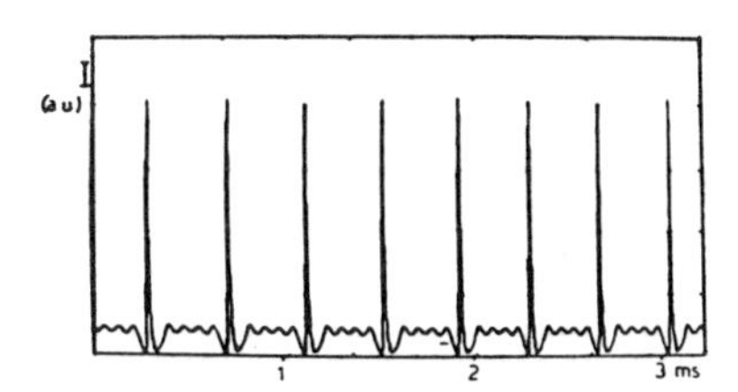

Fig.5 Spiking and regular oscillations in the output intensity vs. time.

injecting amplitude A, the frequency of these pulses goes to zero, because the system remains locked for longer times.

In Fig.6 we show the temporal sequence which leads to chaos, on changing A , by intermittency. Each dot representing the peak of an oscillation , we see, from bottom to top,how the laminar period becomes shorter and shorter and eventually dies in a wholly developed chaos.

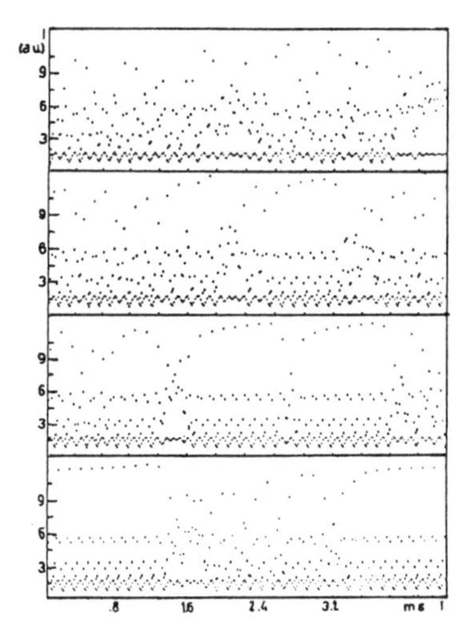

Fig.6 Temporal sequence leading to chaos by intermittency increasing A(from bottom to top),each dot represinting a maximum.

The bifurcation sequence is shown in Fig.7. The bifurcation occurs in the higher spikes while between two near-lyng higher peaks we find still oscillations at Ω .

Last we consider a longitudinal single mode CO_2 ring laser in which both directions of propagation are allowed. The reason why we include this system in the category of single mode lasers in interaction with an external field is the following. The linewidth being homogeneously broadened, the two counterpropagating beams can not work at the same time, because they must compete for the same amount of population inversion. Moreover they are slightly detuned between each other - and with respect to line centre - because, for intrinsic asymmetries, cavity losses are different on the two propagation directions (K_1 and K_2); this results in a different mode pulling and then different lasing frequency. The detuning has been shown experimentally as well as in the numerical solution to be essential for breaking the symmetry between the two directions. A forbidden gap around the center of the molecular line, as well as the interchange of role of forward and backward fields at right and left of the line center are evidence of such a detuning. If $K_{1,2}$ were the "cold" damping rates, they could not differ for reciprocity (in a passive medium thermodynamics forbids such a simmetry breaking). However the $K_{1,2}$ put in eqs. 10 contain the self effect due to the transverse part of the propagation in the medium and account for the finite extent of the laser tube. In fact, the wave equation from which we derive the slowly varying first order equation corresponds to a diffraction pattern with a non linear sources term as

$$P \sim \chi_0 E(1-E^2/E_s^2)$$

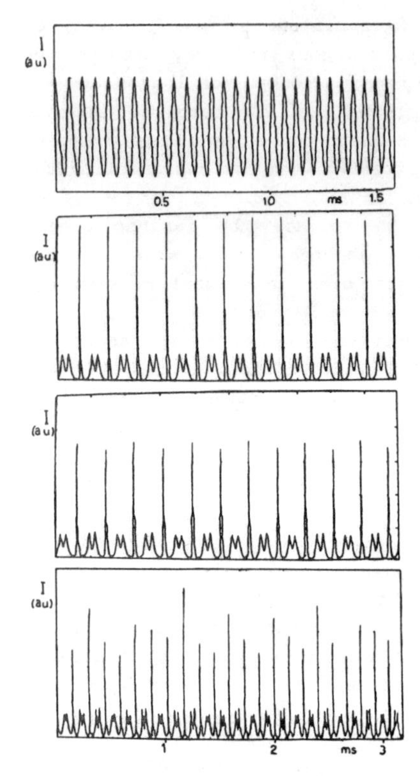

Fig.7 Output intensity vs. time.A is increased from top to bottom and chaos is reached by frequency doubling in the higher peaks.

and already the cubic correction induces a shrinking in the diameter of the beam with a different gain integrated along the cross section. Of course, taking plane waves with infinite lateral extent such an effect would disappear.

Through the grating induced in the population inversion by the interference of the two waves we have an interchange of energy from one field into the other by backscattering. So we obtain again a laser which has a running mode in interaction with a field (the counterpropagating one). Later we will discuss also a configuration as laser with an injected signal.

A modelling of this system is still more complicated. Here together with the detuning from the cavity mode we have to take into account two complex running waves and the induced time dependent grating in the population inversion (truncated at the first order in the expansion). Being x and y the two complex fields, z (real) the spatially uniform component of population inversion and w the complex amplitude of the grating induced in the inversion we have (16)

$$\dot{x} = \frac{1}{1+i\delta}\,[zx + w^{*}y] - x$$

$$\dot{y} = \frac{1}{1+i\delta}\,[zy + wx\,] - \frac{K_2}{K_1}\,y$$

$$\frac{K_1}{\gamma_{/\!/}}\,\dot{z} = (z-z_o) + \frac{1}{1+\delta^2}\,z\,[\,|x|^2+|y|^2] + w^{*}x^{*}y + wxy^{*} \qquad 10$$

$$\frac{K_1}{\gamma_{/\!/}}\,\dot{w} = -w - \frac{1}{1+i\delta^2}\,zx^{*}y + w\,[\,|x|^2+|y|^2+i\delta(|y|^2-|x|^2)]$$

with δ cavity detuning, z_o pump parameter and normalized time $\tau = K_1 t$. Numerical solutions of this seven equation system closely matches all experimental results.

In our parameter space we can distinguish three main different regions showing completely different behaviour. In the first one (Fig.8) we observe a self-pulsing very similar to that of the laser with an injected

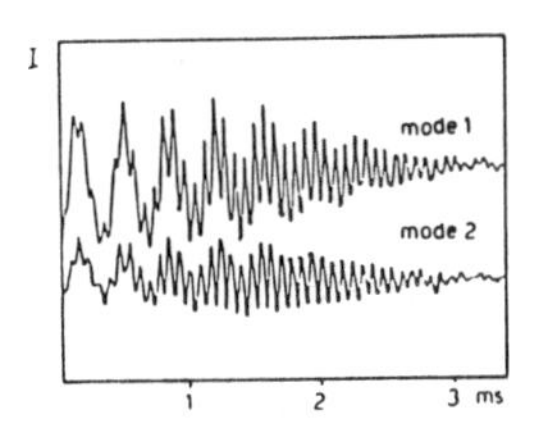

Fig.8 Output intensity vs. time for the two modes.

signal. One mode is also running cw while in the other one we observe only spikes, in phase with the main mode, which occur at a repetition rate (ω_S) of the order of $\gamma_{/\!/}$. In fact, as in the previous system, the cw working mode injects some energy into the other one letting population inversion increase up to a level at which a giant pulse takes place (the height may be 500 times greater than the stationary level). During the pulse both modes go above threshold and spike in phase. Superimposed to the decay we see relaxation oscillations typical of CO_2 lasers with a frequency (ω_0) very near to Ω ; they are out of phase because of competition between the two modes.

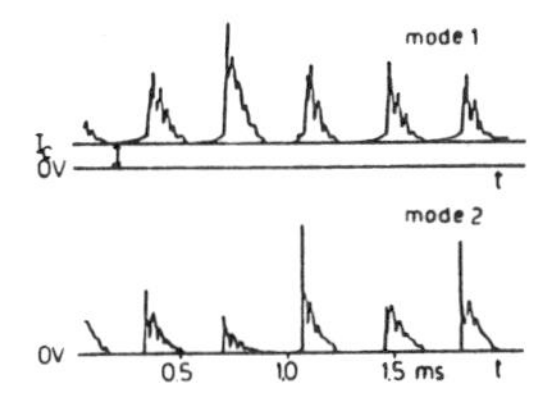

Fig.9 Output intensity vs. time. The upper signal has a cw baseline, while the lower oscillates over a zero level. Oscillations relax to the steady state before a new jump (interchange betxeen cw action) takes place.

For higher excitation currents we observe a deterministic switching due to competition between the two fields with low frequency (~ 30 Hz). During interchange jumps we observe (Fig.9) again the two frequencies of Fig. 8 but with the lower one increased because of a larger value of $\gamma_{/\!/}$ (higher current) while the higher one can be varied also by adjusting the cavity length and alignment by moving a mirror mounted on a piezoelectric crystal.

The transition between these two regimes is not abrupt and it takes place through a region which shows chaotic behaviour. Here both phenomena related to population inversion, spiking (lower currents) and oscillation (higher currents), take place; effective output frequency results also as a combination of the two others (ω_S+ω_0). At the same time if we adjust the cavity mirror position so that we bring $\Omega \sim (\omega_S + \omega_0)$ we obtain a competition of two different variables (population inversion and field) on the same time scale. The result is a fully developed chaos (Fig.10).

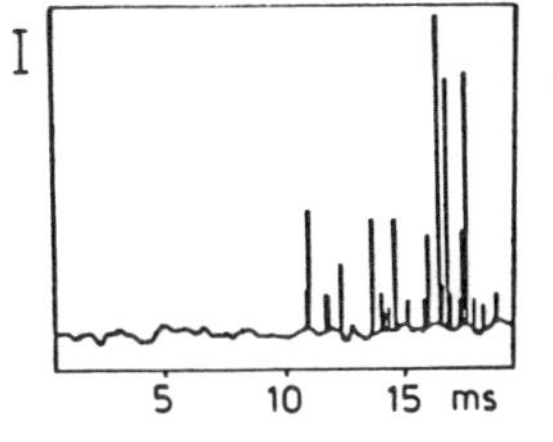

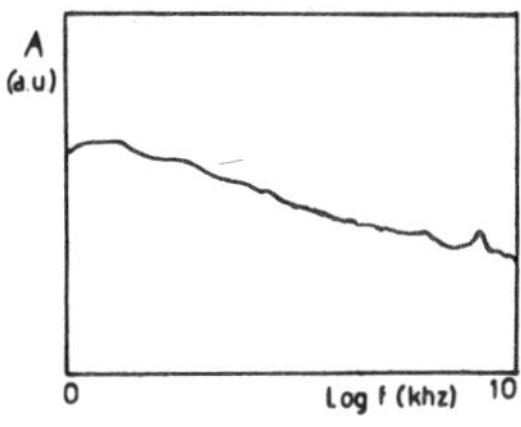

Fig.10 Output intensity vs. time for a wholly chaotic signal (left). Log-log power spectrum with low frequency divergence $f^{-\alpha}$,$\alpha \simeq -0.6$ (right).

If now we inject back one field into the laser with an external mirror we obtain: stabilization of self-spiking, stable laser action instead of switching between the two modes and chaotic behaviour. At the boundary between the spi-

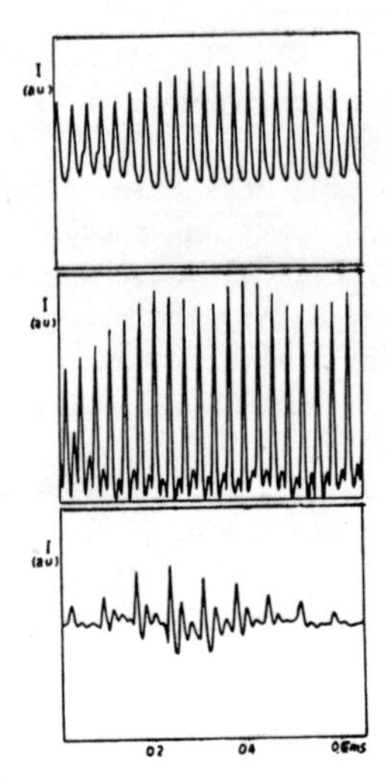

Fig.11 Output intensity vs. time. Bifurcation sequence in analogy with a laser with injected signal.

king and the chaotic region we observe a phenomenology typical of a laser with an injected signal (Fig.11). It means that in this regime we match all parameters which are responsible of such behaviour in the external injection case, although the system here is much more complicated.

Conclusions

We have seen that the single mode CO_2 laser has a rich phenomenology, which we can reproduce numerically with simple theoretical models.

In experiments involving a parameter modulation we obtain much more stable and noise-free outputs so that we can easily compute fractal dimensions and Kolmogorov entropies. At the same time the phenomenology is here not so rich as in experiments where interaction with another field takes place. This must be attributed to the higher complexity of such an interaction where not only an amplitude but also a phase coupling takes place, while in the parameter modulation case interaction is carried only through amplitude modulation.

Summarizing we have found:

i) in a laser with modulated parameters a clear Feigenbaum's route to chaos, with related δ_F and accumulation point evidence

ii) in a laser with an injected signal two different routes to chaos, by intermittency and period doubling

iii) in a bidirectional ring laser self-spiking and chaos .

In this last system we find also a surprising coincidence with a laser with an injected signal when we reflect back one mode into the laser.

We wish to acknowledge help of S. Acciai for mechanical works, L. Albavetti for drawings, S. Mascalchi and P. Poggi for electronics. Furthermore we are indebted to S. Ciliberto, G.L. Oppo, A. Politi and L. Narducci for helpful discussions.

REFERENCES

(a) Also with Dipartimento di Fisica, Università di Firenze.
(b) Permanent address: Bryn Mawr College, Bryn Mawr, PA 19010, USA
(c) Permanent address: Dept. of Chemistry, University of Warsaw,Poland.
(d) Presently working under the CODEST Contract EJOB.
(e) Permanent address: Istituto di Cibernetica CNR Napoli, Italy
(f) Present address: Dept. of Physics, Drexel University, Philadelphia, PA 19104, USA.

(1) F.T. Arecchi XVII Physics Solvay Conference, Bruxelles, 1978 in: "Order and fluctuations in equilibrium and nonequilibrium statistical mechanics", Eds. G. Nicolis et al., J. Wiley 1981.
(2) H. Haken Phys. Lett. 53A, 77(1975).
(3) E.N. Lorenz J. Atmos. Scie. 20, 130(1963).
(4) H. Haken "Synergetics" II ed. Springer (Berlin 1982).
(5)a F.T. Arecchi,G.L. Lippi, G.P. Puccioni and J.R. Tredicce Opt. Comm. 51, 308(1984).
b J.R. Tredicce, F.T. Arecchi, G.L. Lippi and G.P. Puccioni J.Opt.Soc.Am. B2, 173(1985).
(6) A qualitative report on Class C chaos is in
C.O. Weiss, W. Klische, P.S. Ering and M. Cooper Opt. Comm. 52, 408(1985)
however no quantitative measurement of the indicators that characterize chaos has been shown.
(7) F.T. Arecchi, R. Meucci, G.P. Puccioni and J.R. Tredicce Phys. Rev. Lett. 49, 1217(1982).
(8) P. Grassberger and I. Procaccia, Phys. Rev. Lett. 50, 346(1983).
(9) A. Brandstatter, J. Swift, H. Swinney, A. Wolf, J. D. Farmer, E. Jen and J. Crutchfield, Phys. Rev. Lett., 51, 1441(1983).
(10) P. Grassberger and I. Procaccia, Phys. Rev. 28A, 2591(1983).
(11) J.L. Kaplan and J.A. Yorke, Lecture Notes in Math. 730, 228 eds. H.O. Peitgen and H.O. Walther (Springer, Berlin 1980).
(12) P. Grassberger, J. Stat. Phys. 26, 173(1981).
(13) J. P. Crutchfield, J. D. Farmer, B. A. Hubermann, Phys. Rep. 92, 20(1982).
(14) J.R. Tredicce, N.B. Abraham, G.P. Puccioni and F.T. Arecchi Opt. Comm. to appear.
(15) As actually found in accurate experiment on a raser system(radiowave laser) by
E. Brun, B. Derighetti, R. Holzner and D. Meier Helvetica Physica Acta 56, 825(1983).
or in other systems by
W. Klische,H.R. Telle,C.O. Weiss Opt. Lett. 9, 561(1984).
(16) G.L. Lippi, J.R. Tredicce, N.B. Abraham and F.T. Arecchi Opt. Comm. 53, 129(1985).

THE "REDUCED" MODEL THEORY: A SUITABLE THEORETICAL APPROACH FOR DEALING WITH NON-MARKOVIAN, NON-GAUSSIAN STOCHASTIC PROCESSES

Paolo Grigolini

Dipartimento di Fisica e GNSM del CNR, Piazza Torricelli 2, 56100 Pisa Italy

ABSTRACT

The "reduced" model theory is a theoretical procedure which replaces the actual thermal baths with a set of suitable auxiliary variables. The main role of these auxiliary variables is to mimic the non-Gaussian, non-Markovian properties of complex systems without renouncing the description in terms of the standard fluctuation-dissipation processes. This approach is shown to lead to Fokker-Plank equations which are free from the flaws affecting other procedures.

1. Introduction

Several non-linear relaxation processes are currently modelled via the stochastic differential equation

$$\dot{x} = \varphi_1(x) + \varphi_2(x)\, f(t) \qquad 1.1$$

x is the variable of interest and f(t) is a random function of the time t. The main aim of this note is to provide a significant contribution to settling three major problems stemming from this stochastic differential equation. The first problem concerns the choice of the proper algorithm to build up the Fokker-Plank equation to be associated to eq. 1.1 when f(t) is a white noise defined by

$$\langle f(0)f(t)\rangle = 2\, Q\, \delta(t) \qquad 1.2$$

A well known proposal is that of Stratonovich (1), according to the algorithm of whom the probability distribution $\sigma(x;t)$ of the variable x appears to be driven by the Fokker-Plank equation

$$\frac{\partial}{\partial t}\sigma(x;t) = \left[-\frac{\partial}{\partial x}\varphi_1(x) - Q\frac{\partial}{\partial x}\varphi_2(x)\,\varphi_2'(x) + Q\frac{\partial^2}{\partial x^2}\varphi_2^2(x)\right]\sigma(x;t) \qquad 1.3$$

Less popular among physicists is the Itô algorithm (2) which leads to

$$\frac{\partial}{\partial t}\sigma(x;t) = \left[-\frac{\partial}{\partial x}\varphi_1(x) + Q\frac{\partial^2}{\partial x^2}\varphi_2^2(x)\right]\sigma(x;t) \qquad 1.4$$

Note that interesting effects such as noise-induced phase transitions(3-5) are exhibited by the multiplicative stochastic systems, i.e. those described by eq. 1.1 when $\varphi_2'(x) \neq 0$: At a certain critical value Q_c of the parameter Q the shape of the steady state distribution of the variable x undergoes a sudden and significant change (4). On the other hand, eqs. 1.3 and 1.4 predict different critical values (6,7) (it is so precisely because $\varphi_2'(x) \neq 0$) and the Stratonovich algorithm is proven to lead to a satisfactory agreement with the result of the analog simulation (6,7). Section 2 points out the rationale behind this conclusion and shows that physical conditions may exist which provide critical values either closer to the predictions of the Itô algorithm or intermediate between the Itô and the Stratonovich predictions.

A further problem alluded to in Section 2 is the physical interpretation of the instabilities phenomena (8-13) associated with the multiplicative structure of eq. 1.1: The "microscopic" model behind this structure (14) means a steady flow of energy from the environment into the system which is therefore kept far from the state of canonical equilibrium. At certain critical values of this flow (3,15) the system shows a sudden transition from a steady state close to canonical equilibrium to a completely different physical condition, sometimes producing divergent distributions (8-13).

A second major problem discussed in the present paper concerns whether or not the multiplicative term of eq. 1.1 may stem from a purely hamiltonian system. This problem will be dealt with in Section 2.

The third major problem discussed in this paper concerns the case where $\varphi_2(x) = 1$ and f(t) is a Gaussian noise characterized by a finite relaxation time, i.e.,

$$\langle f(0)f(t)\rangle = \langle f^2\rangle e^{-\Gamma t} \quad . \qquad 1.5$$

$\varphi_1(x)$, furthermore, is given the non-linear form

$$\varphi_1(x) = ax - bx^3 . \qquad 1.6$$

Hanggi et al. (16-18) have recently pointed out that some current procedures (1,19) to build up the Fokker-Plank equation to be associated with eq. 1.1 are fraugt with evident flaws. The correct use of the functional derivative method is shown (18) to lead to the non-linear approximate Fokker-Plank equation

$$\frac{\partial}{\partial t}\sigma(x;t) = \left[-\frac{\partial}{\partial x}\varphi_1(x) + \frac{Q}{(1 + (3b\langle x^2\rangle - a)/\Gamma)}\frac{\partial^2}{\partial x^2}\right]\sigma(x;t) . \qquad 1.7$$

The derivation of this Fokker-Plank dynamics is indeed a major result of the recent work of Hanggi et al. (18). In Section 4 I shall show that the "reduced" model theory (RMT) leads to the same result as Hanggi et al.(18)

The RMT is a general strategy to cope with relaxation processes, originally developed to study excitation-relaxation processes of non Markovian quantum-mechanical systems (20-22). The classical-mechanics version (23) of this theory is as follows. Let us consider a set of variables of interest, $\{a\}$. The fluctuation-dissipation nature of these variables stem from the fact that they undergo an uneluctable interaction with a large set of "irrelevant" variables, $\{b\}$. The basic idea of the RMT is to define a suitable subset of variables b_V which fulfills the following major requirement: The "center of gravity" of the system $a + b_V$ is characterized by a time scale slow compared to the remainder of the system. Is is then possible (23) to replace the rigorous Liouville equation describing the time evolution of the total probability distribution $\rho_T(a,b;t)$ with a Fokker-Plank equation driving the reduced description $\rho(a,b\ ;t)$. The intriguing field of the molecular dynamics of the liquid state is precisely where the largest efforts are currently being made to establish rigorous rules to define the set b_V . Eq. 1.1, however, is a case where these auxiliary or "virtual" variables may be established in a completely heuristic way thereby providing a straightforward and intuitive illustration of the RMT.

First of all, let us assume that x is a space coordinate. It is then evident from the fundamental law of the classical mechanics that the stochastic influence of the physical environment must be exerted on the "auxiliary" variable $v \equiv \dot{x}$. Furthermore if the noise f(t) is not white, it will prove necessary to detail the Langevin equation by which this variable is driven.

To deal with all the aforementioned major problems we are led to

$$
\begin{aligned}
\dot{x} &= v \\
\dot{v} &= -\frac{dV}{dx} - \gamma v - \psi'(x)y + F_B(t) \\
\dot{y} &= w \\
\dot{w} &= -\Lambda w - \omega_R^2 y - \varepsilon\psi(x) + F_R(t) \qquad (\varepsilon = 1,0),
\end{aligned}
\tag{1.8}
$$

where $F_B(t)$ and $F_R(t)$ are white noises defined by

$$\langle F_B(0)\ F_B(t)\rangle = 2\gamma k_B T_1 \delta(t) \tag{1.9}$$

$$\langle F_R(0)\ F_R(t)\rangle = 2\Lambda k_B T_2 \delta(t) \tag{1.9'}$$

To start with, let us consider the case when $\varepsilon = 1$. This is a generally non linear stochastic oscillator interacting with a linear one via the interaction term $V_1(x,y) + \psi(x)y$. It is shown (25-28) that this system may be regorously derived from a microscopic Hamiltonian by making a contraction over the infinite degrees of freedom of two independent thermal baths at two different temperatures, T_1 and T_2.

We shall also consider the case when $\varepsilon = 0$. Then the latter stochastic oscillator may be thought of as simulating a radiation field of frequency ω_R, the coherence time duration of which is $1/\lambda$. In this case

T_2 is nothing but a parameter proportional to the intensity of the "radiation" field. Let us assume that

$$\Lambda \gg \omega_R \qquad 1.10$$

This allows us to make the Smoluchowsky assumption $\dot{w} = 0$, which in turn allows us to replace eq. 1.8 with

$$\begin{aligned} \dot{x} &= v \\ \dot{v} &= -\frac{dV}{dx} - \gamma v - \Psi'(x)y + F_B(t) \\ \dot{y} &= -\Gamma y - \frac{\varepsilon}{\Lambda}\Psi(x) + \frac{F_R(t)}{\Lambda} \end{aligned} \qquad 1.11$$

where the fastness of the auxiliary variable y is defined via

$$\Gamma \equiv \omega_R^2/\Lambda \ . \qquad 1.12$$

The limit condition expressed by eq. 1.1 may be approached in different ways. The most general way should rely on putting on the same foot the "irrelevant" variables v and y, which have then to be eliminated simultaneously. A special case of this way is to assume y to be the fastest variable of the system so as to replace eq. 1.11 with equations of motion only concerning the variables x and v. A subsequent contraction over the variable v will lead to a final equation only involving x.

A further special case is to assume γ to be so large as to allow the Smoluchowsky approximation $\dot{v} = 0$, which leads us to replace eq. 1.11 with

$$\begin{aligned} \dot{x} &= -\frac{1}{\gamma}\frac{dV}{dx} - \frac{\Psi'(x)y}{\gamma} + \frac{F_B(t)}{\gamma} \\ \dot{y} &= -\Gamma y - \frac{\varepsilon(x)}{\Lambda} + \frac{F_R(t)}{\Lambda} \ . \end{aligned} \qquad 1.13$$

Then the contraction on the irrelevant variable y can be made. These different ways will be shown in the next Section to lead to different results.

Note that the system subject of the controversy recently settled by Hanggi et al. (16–18) is recovered from eq. 1.13 by making the following assumptions: (i) $\varepsilon = 0$; (ii) $F_B(t) = 0$ (and assuming $\Psi'(x)$ to be independent of x). To recover eq. 1.1 we need the additional Smoluchowsky assumption $\dot{y} = 0$ which leads to

$$\dot{x} = -\frac{1}{\gamma}\frac{dV}{dx} - \frac{\Psi'(x)F_R(t)}{\gamma\Gamma\Lambda} \ . \qquad 1.14$$

Eq. 1.14 has the same structure as eq. 1.1 provided that

$$\varphi_1(x) = -\frac{1}{\gamma}\frac{dV}{dx} \qquad 1.15$$

$$\varphi_2(x) = -\Psi'(x) \qquad 1.16$$

$$f(t) = F_R(t)/\gamma\Gamma\Lambda \quad . \tag{1.17}$$

Before closing this introductory Section, I would like to point out that this paper will be limited to a purely theoretical discussion. Details on the analog experiment corroborating a large part of the theoretical predictions can be found in the companion paper by Leone Fronzoni.

2. A crucial choice: Itô or Stratonovich?

The last remarks of the preceding Section, and especially eq. 1.17 supplemented by eq. 1.9' plus a repeated application of the repartition principle, show that the parameter Q of eq. 1.2 is related to the "reduced" model of eq. 1.8 via

$$Q = \frac{\langle y^2 \rangle}{\Gamma \gamma^2} \tag{2.1}$$

Then we should decide whether eq. 1.3 or eq. 1.4 must be used. We shall follow a different route.

Let us focus our attention on the system of eq. 1.11. The Fokker-Plank equation associated with this set of stochastic equations reads

$$\frac{\partial}{\partial t}\rho(x,v,y;t) = \mathcal{L}\rho(x,v,y;t) \equiv$$

$$\equiv \left\{ -v\frac{\partial}{\partial x} - V'\frac{\partial}{\partial v} + \psi'(x)y\frac{\partial}{\partial v} + \gamma\left[\frac{\partial}{\partial v}v + k_B T_1 \frac{\partial^2}{\partial v^2}\right] + \frac{\varepsilon}{\Lambda}\psi(x)\frac{\partial}{\partial y} + \right. \tag{2.2}$$

$$\left. +\Gamma\left[\frac{\partial}{\partial y}y + \frac{k_B T_2}{\omega_R^2}\frac{\partial^2}{\partial y^2}\right]\right\}\rho(x,v,y;t)$$

The contraction procedure adopted in this paper is that of ref. 23, widely, used in several other papers of the same book (3,24). This technique basically consists of applying the Zwanzig projection method (29) to eq. 2.2 (or, more in general, the effective or rigorous Lìouville equation providing the complete description of the "microscopic" system under study) written in the interaction picture. This means that the most fundamental step of this procedure is the division of the dynamical operator $\mathcal{L}$ into a perturbation, $\mathcal{L}_1$ and an unperturbed part, $\mathcal{L}_0$. This division is strongly dependent on the kind of physical problem we are interested in. For instance, the problem of choosing between the Itô and Stratonovich algorithm certainly refers to a case where both v and y are very fast compared to x. It is therefore sound to assume

$$\mathcal{L}_0 \equiv \gamma\left[\frac{\partial}{\partial v}v + k_B T_1\frac{\partial^2}{\partial v^2}\right] + \Gamma\left[\frac{\partial}{\partial y}y + \frac{k_B T_2}{\omega_R^2}\frac{\partial^2}{\partial y^2}\right] \tag{2.3}$$

$$\mathcal{L}_1 \equiv -v\frac{\partial}{\partial x} - V'(x)\frac{\partial}{\partial v} + \psi'(x)\, y\frac{\partial}{\partial v} \tag{2.4}$$

(we assumed $\varepsilon = 0$). A second-order perturbation calculation along the lines of ref. 23 leads us (by neglecting an irrelevant drift term proportional to γ^{-3} and assuming $T_1 = 0$) to (14,3)

$$\frac{\partial}{\partial t}\sigma(x;t) = \left\{ -\frac{\partial}{\partial x}\varphi_1(x) + Q\left[\frac{\partial}{\partial x}\varphi_2(x)\frac{\partial}{\partial x}\varphi_2(x) + \frac{1}{1+R}\frac{\partial}{\partial x}\varphi_2(x)\varphi_2'(x)\right]\right\}\sigma(x;t), \qquad 2.5$$

where

$$\sigma(x;t) \equiv \int dv\, dy\, \rho(x,v,y;t) \qquad 2.6$$

$$R \equiv \gamma/\Gamma \qquad 2.7$$

We see that eq. 2.5 approachs the Itô or the Stratonovich rule according to whether $R = 0$ or $R = \infty$. Furthermore, it is evident that neither the Itô nor the Stratonovich rule can account for the infinite physical conditions between these two limits.

When $\varphi_1(x)$ is given the analytical form of eq. 1.6, from eq. 2.5 the threshold for the noise-induced phase transition is analytically predicted to depend on R. The analog experiment, widely detailed in the companion paper by Leone Fronzoni, satisfactory corroborated these theoretical predictions (15).

A further intringuing problem completely settled within the context of the RMT (thanks also to a strategy relying on the joint use of analog simulation, theory and computer calculation (30)) is that concerning the relaxation time of x, T, as a function of Q. If T, as claimed by the authors of ref. 19,31 underwent a monotonic increase with increasing Q, the time scale separation between system and bath behind eq. 2.5 would increase, thereby improving the accuracy of this equation. The investigation work of our group (15,32,33) on the contrary, shows that the actual behaviour of T as a function of Q is as follows. After an initial increase, T reaches the maximum at precisely the threshold for the noise-induced phase transition. Then, a further increase of Q makes T decrease till to break the slowness condition on which the contraction over the irrelevant variables y and v relies (15,33). The breakdown of the adiabatic elimination procedure leading to eq. 2.5 is corroborated by the recent discovery (34), via analog simulation, of the oscillatory character displayed in this region by the correlation function $\langle x(0)x(t)\rangle$.

3. The case of hamiltonian systems: Canonical and non-canonical effects.

A satisfactory treatment of the case $\varepsilon = 1$ must cope with the fact that the condition $T_1 = T_2 = T$ implies a canonical equilibrium distribution (25-28) of the whole set of variables x, v and y, which, in turn, leads to the equilibrium state of x described by

$$\sigma_{eq}(x) \propto \exp(-V_{eff}(x)/k_B T) \quad , \qquad 3.1$$

where

$$V_{eff}'(x) = V(x) - \psi^2(x)/2\omega_R^2 . \qquad 3.2$$

To satisfy this constraint we must divide the operator $\mathcal{L}$ of eq. 2.2 into an unperturbed and a perturbation part as follows (25-29)

$$\mathcal{L}_0 \equiv \frac{\varepsilon}{\Lambda} \psi(x) \frac{\partial}{\partial y} + \Gamma\left[\frac{\partial}{\partial y} y + \frac{k_B T_2}{\omega_R^2} \frac{\partial^2}{\partial y^2}\right] \qquad 3.3$$

$$\mathcal{L}_1 \equiv -v\frac{\partial}{\partial x} - V'\frac{\partial}{\partial v} + \gamma\left[\frac{\partial}{\partial v} v + k_B T_1 \frac{\partial^2}{\partial v^2}\right] . \qquad 3.3'$$

This choice is dictated be the need of taking into account those large couplings between (x,v) and y which eventually influence at a large extent the drift of y. The motion of the probability distribution

$$\sigma(x,v;t) \equiv \int dy\, \rho(x,v,y;t) \qquad 3.4$$

is then shown (25,27) to be driven by

$$\frac{\partial}{\partial t}\sigma(x,v;t) = \left\{ -v\frac{\partial}{\partial x} + \gamma\left[\frac{\partial}{\partial v} v + k_B T_1 \frac{\partial^2}{\partial v^2}\right] + V'_{eff}\frac{\partial}{\partial v} + \frac{\langle y^2\rangle_{eq}}{\Gamma} \psi'^2(x) \frac{\partial}{\partial v}\left[\frac{\partial}{\partial v} + \frac{v}{k_B T_2}\right]\right\} \sigma(x,v;t) . \qquad 3.5$$

Note that at $T_1 = T_2 = T$ the constraint of eq. 3.1 is certainly fulfilled by eq. 3.5. From eq. 3.5 we see that the Brownian particle exhibits the effective damping

$$\gamma_{eff} = \gamma + \frac{\langle y^2\rangle}{\Gamma} \frac{\langle \psi'^2(x)\rangle}{k_B T_2} = \gamma + \frac{\langle \psi'^2(x)\rangle}{\Gamma \omega_R^2} , \qquad 3.6$$

the intensity of which increases with increasing $\langle \psi'^2(x)\rangle$. This, in a sense, produces an effect which is the opposite of the transition from the overdamped to the inertial regime of the foregoing Section. The transition from the inertial to the overdamped regime provoked by increasing $\langle \psi'^2(x)\rangle$ has also been observed via analog simulation (25). Eq. 3.5, furthermore, predicts that a sort of phase transition, characterized by a proper slowing down, takes place at a certain critical value of the parameter $\Delta T \equiv T_2 - T_1$. Even this theoretical prediction has been completely corroborated by the results of analog simulation (27).

These results are reminiscent of those of the La Jolla group (35), which derived the Fokker-Planck equation of eq. 3.5 (with $T_1 = T_2 = T$) from the Zwanzig Hamiltonian (36). The approach outlined in this Section leads also to an accurate investigation on the overdamped regime. From eq. 1.13 we obtain indeed ($\varepsilon = 1$)

$$\frac{\partial}{\partial t}\sigma(x;t) = \frac{\partial}{\partial x} \frac{k_B T}{\gamma + \frac{\psi'^2(x)}{\omega_R^2 \Gamma}} \left(\frac{\partial}{\partial x} + \frac{V'_{eff}}{k_B T}\right) \sigma(x;t) , \qquad 3.7$$

which shows that the transition from the inertial to the overdamped regime deviates from the rough prediction

$$\frac{\partial}{\partial t}\sigma(x;t) = \frac{k_B T}{\gamma_{eff}} \frac{\partial}{\partial x}\left(\frac{\partial}{\partial x} + \frac{V'_{eff}}{k_B T}\right)\sigma(x;t) \quad , \qquad 3.8$$

which would stem from eq. 3.6. The accurate computer calculations of ref. 25 confirm that eq. 3.7 is the proper description of the transition from the inertial to the overdamped regime. It should be stressed that eq. 3.7 holds even if the system were overdamped with $\psi'^2(x) = 0$.

Furthermore, eq. 3.7 provides a resummation at infinite order on the perturbation $\psi'^2(x)$, thereby affording an effective way, alternative to that followed by Lugiato et al. (37), to solve this intriguing problem.

Before closing this Section, I would like to point out that at $T_2 = T_1$ the final form of the reduced equation, eq. 3.7, does not depend on the road followed (an initial contraction over y followed by the elimination of v, or viceversa). On the contrary, the multiplicative structure of eq. 2.5 (ranging from the Itô to the Stratonovich form with increasing R from 0 to ∞) reappears when $T_2 > T_1$ (as an additional contribution to the r.h.s. of eq. 3.7). This means that such multiplicative structure is associated with an energy pumping from the environment into the system. The relevance of this phenomenon in the field of enzyme chemistry has been discussed in ref. 28.

4. The prblem of a thermal bath with a finite correlation time.

Let us now focus our attention on the eqs. 1.8 for $\varepsilon = 0$, and $\psi'(x) = \omega_I^2$. This is a Brownian anharmonic oscillator under the influence of a radiation field. When $\Lambda \ll \omega_R$ the radiation field is selective enough as to produce resonant activation of the system (in the inertial case). The corresponding Fokker-Planck equation reads

$$\frac{\partial}{\partial t}\rho(x,v,y,w;t) = \mathcal{L}\rho(x,v,y,w;t) =$$

$$= \left\{\left[-v\frac{\partial}{\partial x} + V'\frac{\partial}{\partial v} + \gamma\left(\frac{\partial}{\partial v}v + k_B T\frac{\partial^2}{\partial v^2}\right)\right] + \omega_I^2\frac{\partial}{\partial v}y + \right. \qquad 4.1$$

$$\left. + \left[-w\frac{\partial}{\partial y} + \omega_R^2 y\frac{\partial}{\partial w} + \Lambda\left(\frac{\partial}{\partial w}w + \langle w^2\rangle_{eq}\frac{\partial^2}{\partial w^2}\right)\right]\right\}\rho(x,v,y,w;t)$$

The major aim of this Section is to carry out the contraction of the freedom degrees y and w appearing in eq. 4.1.

This contraction procedure requires especial caution. Let us call $1/\tau_R$ the largest of the two parameters λ and ω_R , that is,

$$1/\tau_R \equiv \mathrm{Max}\,(\Lambda\,,\,\omega_R) \quad . \qquad 4.2$$

Let us also assume that the anharmonic term of V(x) is so weak as to make it possible to characterize the dynamics of the system of interest via the parameters γ and ω_{eff} (where ω_{eff} is an effective frequency to be defined later). Then we define

$$1/\tau_B \equiv \mathrm{Max}\,(\gamma, \omega_{eff}) \quad . \tag{4.2'}$$

If

$$\tau_R \ll \tau_B \tag{4.3}$$

we are allowed to choose tha last term between square brackets of eq. 4.1 as the unperturbed part of the operator $\mathcal{L}$.

The photoselective phenomena certainly invalidate the condition 4.3. Another case where this condition is not fulfilled is that studied by Hanggi et al. (16-18). To study these physical conditions we must use the following division of $\mathcal{L}$:

$$\begin{aligned}
\mathcal{L} &= \mathcal{L}_0 + \mathcal{L}_1 \\
\mathcal{L}_0 &\equiv \mathcal{L}_a + \mathcal{L}_b \\
\mathcal{L}_a &\equiv -v\frac{\partial}{\partial x} + V'\frac{\partial}{\partial v} + \gamma\Big(\frac{\partial}{\partial v}v + k_B T\frac{\partial^2}{\partial v^2}\Big) \\
\mathcal{L}_b &\equiv -w\frac{\partial}{\partial y} + \omega_R^2\frac{\partial}{\partial w} + \Lambda\Big(\frac{\partial}{\partial w}w + \langle w^2\rangle\frac{\partial^2}{\partial w^2}\Big) \\
\mathcal{L}_1 &\equiv y\,\omega_I^2\frac{\partial}{\partial v}
\end{aligned} \tag{4.4}$$

Let us call $\rho_{eq}(y,w)$ the radiation field equilibrium distribution, i.e.,

$$\mathcal{L}_b\,\rho_{eq}(y,w) = 0 \quad . \tag{4.5}$$

The projection operator P to be used is then

$$P\,\rho(x,v,y,w;t) \equiv \rho_{eq}(y,w)\int dy\,dw\,\rho(x,v,y,w;t) \quad . \tag{4.6}$$

By applying the usual procedure (23) we obtain

$$\begin{aligned}
\frac{\partial}{\partial t}\sigma(x,v;t) &= \mathcal{L}_a\,\sigma(x,v;t) + \\
&+ \frac{\omega_I^2}{\rho_{eq}(y,w)}\,P\int_0^t d\tau\, y\frac{\partial}{\partial v}\exp(\mathcal{L}_b + \mathcal{L}_a)(t-\tau))\frac{\partial}{\partial v}\sigma(x,v;t)\,\rho_{eq}(y,w).
\end{aligned} \tag{4.7}$$

We must now make $\exp\{\mathcal{L}_a(t-\tau)\}$ commute with $\partial/\partial v$. Note that

$$\Big[\mathcal{L}_a, \frac{\partial}{\partial v}\Big] = \frac{\partial}{\partial x} - \gamma\frac{\partial}{\partial v} \quad , \tag{4.8}$$

$$\left[\mathcal{L}_a , \frac{\partial}{\partial x} \right] = \alpha \frac{\partial}{\partial v} - 3 x^2 \beta \frac{\partial}{\partial v} . \qquad 4.9$$

The last term on the r.h.s. of eq. 4.7 means that by iterative application of the superoperator $\mathcal{L}_a^{\times} \equiv [\mathcal{L}_a, \cdots]$, an infinite chain is generated. This hierarchy is truncated at the second step by the assumption

$$\left[\mathcal{L}_a , \frac{\partial}{\partial x} \right] = -\omega_{eff}^2 \frac{\partial}{\partial v} \equiv - (3 \langle x^2 \rangle \beta - \alpha) \frac{\partial}{\partial v} . \qquad 4.10$$

The effective frequency ω_{eff} defined by eq. 4.10 coincides with the one resulting in the correction term of ref. 18 (note that ref. 18 deals with $ax - bx^3$ which, in the overdamped regime, comes from $(\alpha x - \beta x^3)/\gamma$). More in general, the aforementioned infinite hierarchy results in an expansion with derivatives of any order. When assuming that throughout its motion $\sigma(x,v;t)$ undergoes only weak deviations from its equilibrium state, the diffusion coefficient, i.e., the strength of the term $\partial^2/\partial v^2$, is shown to coincide with the theoretical predictions of the linear response theory (38).

This is precisely the way currently followed to deal with the escape processes from a potential well in highly resonant systems in the presence of radiative excitation. It is interesting to remark that the system studied by Hanggi et al. (18) is recovered with eqs. 1.8 by considering overdamped oscillators (and disregarding $F_B(t)$). The approach outlined in the present Section leads then to eq. 1.7 thereby corroborating the point of view of Hanggi et al. (18).

The case of extremely underdamped non-linear systems is a puzzling problem by itself (i.e. even in the absence of the radiation field). The major result, achieved by the strategy outlined in this paper, is the discovery of the breakdown of the renormalization techniques relying on the statistical linearization method (40), which, therefore, has to be replaced by an alternative renormalization procedure (41), which, still, leads to analytical predictions. These have been satisfactorily corroborated by the results of analog simulations (41).

As a final remark, we would like to stress that our conclusions on the choice between the Itô and the Stratonovich rule agrees with those of other authors (42). The investigation on the low-friction limit has been stimulated by the recent theoretical investigation of Evans (43).

Acknowledgment

I warmly thank Drs. S. Faetti, T. Fonseca, D. Pareo, B. Zambon and Mr. R. Mannella who gave significant contribution to a wide part of the theoretical work here reviewed.

References

1 R.L. Stratonovich, Topics in the Theory of Random Noise, Gordon and Breach, New York, Vol. 1, 1963, Vol. 2, 1967.

2 K. Itô, Mem. Math. Soc., 4, 1 (1961).

3 S. Faetti, C. Festa, L. Fronzoni and P. Grigolini, in "Memory Function Approaches to Stochastic Problems in Condensed Matter", Special Issue of Advances in Chemical Physics, eds. M.W. Evans, P. Grigolini, G. Pastori Parravicini, general editors I. Prigogine and S.A. Rice, Vol. 62 (to appear on July 1985).

4 Schenzle and M. Brand, Phys. Rev. 20A, 1628 (1979).

5 W. Horshemke and R. Lefever, Noise Induced Transitions (Springer-Verlag Berlin, 1984).

6 I. Smythe, F. Moss, and P.V.E. McClintock, Phys. Lett. 97A, 95 (1983).

7 I. Smythe, F. Moss, and P.V.E. McClintock, Phys. Rev. Lett. 51, 1062 (1983).

8 B.J. West, K. Lindenberg, and V. Seshadri, Physica, 102A, 470 (1980).

9 K. Lindenberg, V. Seshadri, K.E. Shuler and B.J. West, J. Stat. Phys. 23, 755 (1980).

10 K. Lindenberg, V. Seshadri, and B.J. West, Phys. Rev. 22, 2171 (1980).

11 N.G. Van Kampen, Physica, 102A, 489 (1980).

12 P. Hanggi, Phys. Lett. 78A, 489 (1980).

13 P. Grigolini, Phys. Lett. 84A 301 (1981).

14 S. Faetti, P. Grigolini, and F. Marchesoni, Z. Phys. 47B, 535 (1982).

15 S. Faetti, C. Festa, L. Fronzoni, P. Grigolini and P. Martano, Phys. Rev. 30A, 3252 (1984).

16 P. Hanggi, F. Marchesoni, and P. Grigolini, Z. Phys. 56B, 333 (1984).

17 P. Hanggi, Z. Phys. B31, 407 (1978); P. Hanggi, "The Functional Derivative and its Use in the Description of Noisy Dynamical Systems", in: Stochastic Processes Applied to Physics, eds. L. Pesquera and M. Rodriguez, (World Scientific, 1985).

18 P. Hanggi, T.J. Mroczkowski, F. Moss and P.V.E. McClintok, preprint (1985).

19 J.M. Sancho, M. San Miguel, S.L. Katz and J.D. Gunton, Phys. Rev. 26A, 1589 (1982); M. Lax, Rev. Mod. Phys. 38, 541 (1966); N.G. van Kampen, Phys. Rev. 24C, 171 (1976).

20 P. Grigolini, Chem. Phys. Lett. 47, 483 (1977).

21 P. Grigolini and A. Lami, Chem. Phys. 30, 61 (1978).

22 P. Grigolini, Il Nuovo Cimento 63B, 17 (1981).

23 P. Grigolini, in "Memory Function Approaches to Stochastic Problems in Condensed Matter", Special Issue of Advances in Chemical Physics, eds. M.W. Evans, P. Grigolini, G. Pastori Parravicini, general editors I. Prigogine and S.A. Rice, Vol. 62 (to appear on July 1985).

24 M. Ferrario, P. Grigolini, A. Tani, R. Vallauri, and B. Zambon, in "Memory Function Approaches to Stochastic Problems in Condensed Matter", Special Issue of Advances in Chemical Physics, eds. M.W. Evans, P. Grigolini, G. Pastori Parravicini, General eds. I. Prigogine and S.A. Rice, Vol. 62 (to appear on July 1985); M.W. Evans, ibid.

25 S. Faetti, L. Fronzoni, P. Grigolini, Phys. Rev. A (in press).
26 T. Fonseca, P. Grigolini, D. Pareo, J. Chem. Phys. (in press).
27 L. Fronzoni, P. Grigolini, Phys. Lett. 106A, 289 (1984).
28 M. Compiani, T. Fonseca, P. Grigolini, R. Serra, Chem. Phys. Lett. (in press).
29 R. Zwanzig, J. Chem. Phys. 33, 1338 (1960); Lectures in Theoretical Physics, Vol. III, W.E. Brittin, B.W. Downs, and J. Downs, eds., Interscience, New York, 1961 pp. 106-141.
30 The computer calculation used by our group is basically a continued fraction procedure relying on the same theoretical background as the RMT itself (see ref. 23). See also: G. Grosso, G. Pastori-Parravicini (ibid).
31 A. Hernandez-Machado, M. San Miguel and J.M. Sancho, Phys. Rev. 29A, 3388 (1984).
32 S. Faetti, C. Festa, L. Fronzoni, P. Grigolini, F. Marchesoni, and V. Palleschi, Phys. Lett. 99A, 25 (1983).
33 C. Festa, L. Fronzoni, P. Grigolini, F. Marchesoni, Phys. Lett. 102A, 95 (1984).
34 L. Fronzoni, P. Grigolini, P. Martano, submitted to Phys. Rev. A.
35 K. Lindenberg, V. Seshadri, Physica 109A, 483 (1981).
36 R. Zwanzig, J. Stat. Phys. 9, 215 (1973).
37 L.A. Lugiato, Physica 81A, 565 (1976); F. Casagrande, E. Eschenazi, and L.A. Lugiato, Phys. Rev. 29A, 239 (1984); L.A. Lugiato, P. Mandel, L.M. Narducci, Phys. Rev. 29A, 1438 (1984).
38 T. Fonseca, P. Grigolini, submitted to Phys. Rev. A.
39 L. Fronzoni, P. Grigolini, R. Mannella, and B. Zambon, Phys. Lett. 107A, 204 (1985).
40 A.B. Budgor, J. Stat. Phys. 15, 355 (1976); A.B. Budgor, K. Lindenberg, and K.E. Shuler, J. Stat. Phys. 15, 375 (1976); A.B. Budgor, B.J. West, Phys. Rev. 17A, 370 (1978).
41 L. Fronzoni, P. Grigolini, R. Mannella, B. Zambon, submitted to J. Stat. Phys.
42 R. Graham and A. Schenzle, Phys. Rev. 26A, 1676 (1982); C.W. Gardiner, Phys. Rev. 29A, 2814 (1984); C.W. Gardiner, M.L. Steyn Ross, Phys. Rev. 29A, 2823 (1984).
43 M.W. Evans, Molec. Phys. 53, 1285 (1984), Physica Scripta 30, 222 (1984).

ANALOG SIMULATION OF NONLINEAR STOCHASTIC PROCESSES BY MEANS OF ELECTRIC DEVICES

Leone Fronzoni
Dipartimento di Fisica e GNSM
Piazza Torricelli 2, 56100 Pisa - Italy

ABSTRACT

The analog simulation by means of electric circuits is an efficient tool to study nonlinear stochastic processes. This method is proved to afford a reliable check of the current theories, while also suggesting the way to new theoretical modelling. In this paper, the foundations of the analog simulation are presented and some significant examples are given in the spirit of the reduced model theory.

Introduction

The study of nonlinear stochastic processes, especially far from the equilibrium state, has recently attracted considerable attention. The solutions of these problems is usually fraught with large difficulties mainly stemming from the nonlinear character of these phenomena. This means that the treatment of these processes is widely beyond the range of analytical methods and makes it necessary the recourse to a computer calculation. However, this often implies the waste of a large amount of time and, therefore, high costs.

Analog simulation affords a straightforward solution to these difficulties, though involving less precision than computer calculation. It also allows us to reach the results in real time, thereby allowing for a detailed description of the dynamics (1-14).

In our laboratory it has been recently developed an analog simulation technique in order to study a wide class of stochastic phenomena. In this paper we do not pay a deep attention to the theoretical aspects which are analysed in the companion paper by Paolo Grigolini.

In Section I we give a short description of the basic ideas behind analog simulation and noise generator.

In Section II we present the results of the experiment on a nonlinear system driven by multiplicative stochastic forces.

In Section III we describe an experiment on two oscillators driven by two independent fluctuation-dissipation processes, and coupled via an hamiltonian interaction.

In Section IV an interesting picture of a Duffing oscillator in the low-friction limit is given.

Section I

A large variety of physical problems is described by a set of differential equations with the form

$$\dot{x}_i = a_{ij} x_j + b_{ijk} x_j x_k + \xi_i(t) \qquad 1$$

where $\xi_i(t)$ are stochastic variables. Eqs. 1 can be simply modelled by means of electric circuits, resorting to the following basic operations:

----a) Integration

The RC-circuit (Fig. 1a) is well known to be an integrator. Indeed, from the relation

$$(RC) \cdot V_u = \int (V_i - V_u) dt \qquad 1a$$

we see that the output signal V_u is an integral function of the input signal V_i. This circuit presents, however, the inconvenience that, when coupled with other circuits, the value of the integration parameter changes. The Miller integrator overcomes this inconvenience, as it relies on a differential amplifier with a capacitor feed-back function (Fig. 1b). The electric property of this element, high input impedence and low output impedence, allows the relation between V_i and V_u to be written in the differential form

$$V_i = - RC \, \dot{V}_u \qquad 1b$$

----b) Sum

This operation is feasible using the same basic principle as that behind the afore illustrated device (Fig. 1c)

$$V_u = \frac{R}{R_1} V_1 + \frac{R}{R_2} V_2 \qquad 1c$$

----c) Product

We have carried out the product operation by means of a four quadrant multiplier (Analog Device AD534) which presents three differential inputs (x,y,z). The transfer function is

$$V_u = \frac{(x_1 - x_2)(y_1 - y_2)}{10} + (z_1 - z_2) \qquad 1d$$

The three differential inputs allowed us to perform simultaneously the sum and the product operation with a high degree of flexibility.

----d) Noise generator

The simulation of stochastic processes requires a highly accurate generation of white Gaussian noise. This is made possible by applying the operating principle of a linear-feedback shift register (LFSR) supplemented by significant improvements. For additionals details the reader can refer to a previous paper (25). Anyway the scheme of the electric circuit is shown in Fig. 2.

The 17-stages LFSR, with taps at stages 5 and 17 of the shift register SR1, is driven by the stage 4 of the flip-flop FF and, in turn, drives the main 17-stages LFSR, which also works with taps at 5 and 17 of the shift-register SR2. A low-pass filter gives a cut-off frequency $\nu_T = 1/R_2C_2$, thus transforming the telegraphic noise in Gaussian noise.

We have proved that the noise is Gaussian if the cutoff frequency is lower than 1/50 of the clock frequency. Moreover, from the Fourier analysis, the noise spectrum is proven to be flat below the cut-off frequency.

Fig. 3 shows the distribution of the probability density versus the noise amplitude. The empty triangles refer to the case when only one LSFR is used, and we see how the curve is affected by skewness. A much higher accuracy is reached when using two LSFR (full triangle curve).

SECTION II

The dynamics of many stochastic processes, is ruled by a differential equation (15) as

$$\dot{v} = -\gamma v + V'(x) + \varphi(x)\xi(t) \qquad 2$$

where V(x) is a nonlinear potential and $\xi(t)$ is a white and Gaussian noise

$$\langle \xi(o)\ \xi(t)\rangle = 2Q\,\delta(t)$$

Because of the presence of the multiplicative term $\varphi(x)\xi(t)$, it is not univocally defined which Fokker-Planck is to be associated with eq. 2. Two algorithms are, at least,possible, depending on the choice of the Itô or Stratonovich representation (1,6).

In the companion paper, Grigolini shows that this ambiguity is removed if one replaces eq. 2 with a new set of equations

$$\begin{aligned} \dot{x} &= v \\ \dot{v} &= -\gamma v + V'(x) + \varphi(x)\xi(t) \\ \dot{\xi} &= -\lambda\xi + f_s(t) \end{aligned} \qquad 3$$

where $\langle f_s(0)\ f_s(t)\rangle = 2D\ \delta(t)$, and the ambiguity is removed being the stochastic variable represented by an additive term. Theoretical considerations (25) suggest that in the limit case $\lambda, \gamma \to \infty$, the solution is heavily dependent on the ratio $r = \gamma/\lambda$ which leads the system either to the Itô ($r \to 0$) or to the Stratonovich regime ($r \to \infty$). We have tested this prediction with an analog circuit along the lines describes in Section I.

Let us refer for simplicity to the potential V(x)

$$V(x) = \frac{1}{2}\alpha x^2 + \frac{1}{4}\beta x^4 + \sigma_0$$

and $\varphi(x) = x$. Fig. 4 shows the scheme of the electric circuit. The equations written at the nodes of the circuit give

$$\dot{V}_2 = -\frac{R_1 + R_2}{R_1 C_1} V_2 + \frac{V_{eq}^2}{R_1 R_2 C_1 C_2 V_r^2} V_2 - \frac{V_2^3}{R_1 R_2 C_1 C_2} + \frac{\xi(t) V_2}{R_1 R_2 C_1 C_2} \quad 4$$

$$\ddot{x} = -\gamma \dot{x} + \alpha x - \beta x^3 + x \xi(t) , \quad 5$$

while at the low-pass filter after the noise injection, we have

$$\dot{\xi} = -\lambda\xi + f_s(t)$$

This and the previous equation provide an equivalent picture to that given by the set 3.

By applying the Adiabatic Elimination Procedure (17,18) and considering x the slow variables, we obtain the Fokker-Planck equation

$$\dot{\sigma} = \left[-\frac{\partial}{\partial x} (dx - bx^3) + Q \frac{\partial}{\partial x} x \frac{\partial}{\partial x} \right] \sigma \quad 6$$

where $d = \frac{do}{\gamma} - \frac{Q}{1+r}$; $do = \frac{d}{\gamma}$; $Q = \frac{\langle f^2 \rangle}{\gamma^2 \lambda}$; $r = \gamma/\lambda$; $b = \frac{\beta}{\gamma}$

Two critical values of the parameter Q exist

$$Q_1 = do \frac{1 + r}{2 + r} , \qquad Q = do\,(1 + r)$$

Q_1 corresponds to a transition for the probability density from a Gaussian-like shape to a divergent one around $x = 0$ (see Fig. 5). Q_2 corresponds to the onset of a null second moment ($\langle x^2 \rangle_{ss} = 0$). In the weak noise region the experimental results agree with the theoretical prediction. Fig. 6 confirms the transition from the Itô to the Stratonovich regime, when r is changed. On the contrary, in the large Q region, $\langle x^2 \rangle$ not only does not vanish, but, after having reached a minimum value, it begins to increase with increasing Q (see Fig. 7).

A direct measurement of the correlation function $\langle x(0)x(t) \rangle$ shows a transition from the overdamped to the inertial regime, where an oscillating behaviour is found (see Fig. 8). It suggests that, for large noise-values, x(t) becomes a fast variable thereby invalidating the AEP. According to Seshadri et al. (19), the energy $E = \langle v^2 \rangle/2 + \omega_0^2 \langle x^2 \rangle /2$ can be regarded as a new slow variable. As a consequence, $\langle x^2 \rangle$ turns out to be proportional to Q, thus in agreement with the analog results shown in Fig. 7.

SECTION III

Let us now consider the set of equations

$$\dot{x} = v$$

$$\dot{v} = -\gamma v - \omega_0'^2 - \beta x^3 - 2\alpha yx + f_1(t) \quad 7a$$

$$\dot{y} = w$$

$$\dot{w} = -\lambda w - \omega_0^2 y - \alpha x^2 + f_2(t) \qquad 7b$$

Where $f_1(t)$ and $f_2(t)$ are white and Gaussian noise defined by

$$\langle f_1(0)f_1(t)\rangle = 2Q_1\,\delta(t) \qquad \langle f_2(0)f_2(t)\rangle = 2Q_2\,\delta(t)$$

Eqs. 7 describe the motion of two coupled oscillators, interacting with two independent thermal baths. The coupling term originates from the potential energy term $V(x,y) = \alpha\, yx^2$.

Fig. 9 shows the experimental configuration simulating eqs. 7. The circuit of the anharmonic oscillator is the same as that in Section II. A detailed study has been performed for different values of the temperatures $T_1 = Q_1/\gamma k_B$ and $T_2 = Q_2/\lambda k_B$, with k_B being the Boltzmann constant. The value of the thermal energy $k_B T$ can be easily evaluated by exploiting the relation $\langle v^2\rangle_{eq} = k_B T$. Indeed, $\langle v^2\rangle$ is measured by sending the v-signal to an analog multiplier and then to a low-pass filter.

Canonical case: $T_1 = T_2$

By recalling the relations

$$\langle v_x^2\rangle = k_B T_1 \qquad \langle v_y^2\rangle = k_B T_2$$

it is immediately seen that the condition $T_1 = T_2$ is easily achieved by setting $\langle v_x^2\rangle\ \langle v_y^2\rangle$.

In our experiments we have chosen $w_0 = 10\ w_0'$, $\gamma = .3\omega_0'$ and $\Gamma = \frac{\omega_0^2}{\lambda} = 3\omega_0'$, these values verifying the inequality

$$\Gamma \gg \omega_0' \gg \gamma$$

This condition, in turn, allows to test a theoretical relation based on the mean-field approximation (20). The experimental values of the effective friction coefficient γ_{eff} are displayed in Fig. 10 (full circles) as a function of the thermal energy $k_B T$. They are to be compared with the theoretical relation (continuous line)

$$\gamma_{eff} = \gamma + \alpha^2\langle v^2\rangle/\Gamma\omega_0^2$$

The experimental data have been obtained by fitting the correlation function

$$\phi(x) = \frac{\langle x(0)x(t)\rangle - \langle x^2\rangle}{\langle x^2\rangle - \langle x\rangle^2}$$

Technical difficulties occur in the viscous regime ($\gamma > \omega_0'$) because a high intensity noise is necessary for obtaining an appreciable variation of γ_{eff}.

Non-canonical case

In the inertial regime the proper slow variable is the energy $E = v^2/2 + \omega_0' x^2/2$. By following the Stratonovich theoretical approach, widely used in a recent paper by Lindenberg at al. (21), we see that the Langevin equations 7 yield

$$\langle E(t)\rangle = \frac{C(E_0 - C_2) - C_2(E_0 - C_1)\exp(-\eta t)}{(E_0 - C_2) - (E_0 - C_1)\exp(-\eta t)} \qquad 8$$

where

$$C_{1,2} = -\frac{1}{2}\left[\left(\frac{\lambda_0 \omega_0'^2}{\lambda_2} - k_B T_2\right) \pm \frac{1}{2}\left(\frac{\lambda_0 \omega_0'^2}{\lambda_2} - k_B T_2\right) + 4\omega_0' k_B T_1 \frac{\lambda_0}{\lambda_2}\right]^{1/2}$$

$$\eta \equiv \left[\left(\lambda_0 - \frac{k_B T_2}{\omega_0'^2}\lambda_2\right)^2 + 4 k_B \frac{T_1 \lambda_2 \lambda_0}{\omega_0'^2}\right]^{1/2}, \quad \lambda_2 = \frac{\alpha^2}{\omega_0^2 \Gamma} \qquad \lambda_0 = \gamma$$

By assuming T_1 to vanish, $\langle E(\infty)\rangle$ turns out to be

$$\langle E(\infty)\rangle = 0 \qquad 0 \leq T_2 \leq T_c$$
$$\langle E(\infty)\rangle = k_B(T_2 - T_c) \qquad T_2 > T_c,$$

where the threshold T_c is defined by

$$T_c = \lambda_0 \omega_0'^2 / \lambda_2 k_B$$

The condition $T_1 \simeq 0$ means that the amplitude of the noise applied to the nonlinear oscillator vanishes. Fig. 11 shows the experimental results, where $\langle x^2\rangle$ is plotted as a function of $\langle y^2\rangle = k_B T_2/\omega_0^2$. The ideal case $T_1 = 0$ cannot be experimentally reproduced because of the internal noise of the electric circuit. However, the intensity of the spurious noise decreases with decreasing $\langle y^2\rangle$ and in the weak $\langle y^2\rangle$ region the internal noise is shown to almost vanish (see Fig. 11).

SECTION IV - The Duffing oscillator in the low-friction regime.

Let as consider the stochastic system described by

$$\dot{x} = v$$
$$\dot{v} = -\gamma v - \omega_0^2 x - \beta x^3 + f_S(t) \qquad 9$$

where $f_S(t)$ denotes a white and Gaussian noise defined by

$$\langle f_S(0) f_S(t)\rangle = 2\gamma k_B T \delta(t)$$

Theoretical remarks (23) lead to a temperature-dependent characteristic frequency

$$\alpha = \frac{3}{4}\beta\, k_B T/\omega_0^3$$

In the low-friction limit ($\gamma \ll \alpha$) the x-spectrum of the system should exhibit a maximum at $\omega = \omega_0 + \alpha$, while for $\gamma \gg \alpha$ the maximum should be found at $\omega = \omega_0 + 2\alpha$. We have assessed the validity of these theoretical predictions. The experimental device is basically similar to that described in the previous Section. A scheme of the apparatus is shown in Fig. 12. The damping γ was varied by changing a feed-back resistence in the first integrator.

The extremely low-friction regime involves technical difficulties such as an high sensibility to the internal noise and a thermal drift affecting the long-time averages. As a conclusive illustration of our technique we show in Fig. 13 the spectrum for several values of γ. Furthermore, in Fig. 14 the experimental results are compared with the theoretical prediction (24) for $\gamma \ll \alpha$.

Acknowledgement

I am specially grateful to professor Paolo Grigolini for his advice, comments and a critical reading of the manuscript.

REFERENCES

1 - R.L. Stratonovich, Topics in the theory of Random Noise; vol. 1 (Gordon and Breach, New York 1967).
2 - R. Landauer, J. Appl. Phys. 33, 2209 (1962).
3 - K. Matsuno, Appl. Phys. Lett. 12, 404 (1968).
4 - James W.F. Woo and R. Landauer, IEEE J. Quantum El. 7, 435 (1971).
5 - T. Kawakubo, S. Kabashima and K. Nishimura, J. Phys. Soc. Jpn 34, 1940 (1973).
6 - S. Drosdziok, Z. Physik 261, 431 (1973).
7 - J.B. Morton and S. Corrins, J. Math. Phys. 10, 361 (1969).
8 - P.M. Horn, T. Carruthers and M.T. Long, Phys. Rev. A 14, 833 (1976).
9 - S. Kabashima, S. Kogure, T. Kawakubo and T. Okada, J. Appl. Phys. 50, 6296 (1979).
10 - Y. Morimoto, J. Phys. Soc. Jpn, 52, 1086 (1983).
11 - K. Harada, S. Kuhmura and K. Hirakawa, J. Phys. Soc. Jpn 50, 2450 (1981).
12 - J.M. Sancho, M. San Miguel, H. Yamzaki and T. Kawakubo, Physica 116A, 560 (1982).
13 - I. Smythe, F. Moss, P.V.E. Mc Clintock, Phys. Lett. 97A, 95 (1983).
14 - I. Smythe, F. Moss, P.V.E. Mc Clintock, Phys. Rev. Lett. 51, 1062 (1983).
15 - R. Graham and A. Schenzle, Phys. Rev. A 26, 1676 (1982).
16 - K. Itô, Men. Am. Math. Soc. 4, 1 (1951).
17 - S. Faetti, P. Grigolini and F. Marchesoni, Z. Phys. B47, 535 (1982).
18 - P. Grigolini and F. Marchesoni, Adv. Chem. Phys. (in press).
19 - V. Seshadri, B.J. West and K. Lindenberg, Physica (Utrecht) A107, 289 (1981).
20 - S. Faetti, L. Fronzoni and P. Grigolini, Phys. Rev. (in press).

2£ - K. Lindenberg and V. Seshadri, Physica 109A, (1981).
22 - L. Fronzoni and P. Grigolini, Phys. Lett. 106A, 289 (1984).
23 - L. Fronzoni, P. Grigolini, R. Mannella and B. Zambon, Phys. Lett. 107A, 204 (1985).
24 - L. Fronzoni, P. Grigolini, R. Mannella and B. Zambon, submitted to J. Stat. Phys.
25 - S. Faetti, C. Festa, L. Fronzoni, P. Grigolini and P. Martano, Phys. Rev. A30, 3252 (1985).

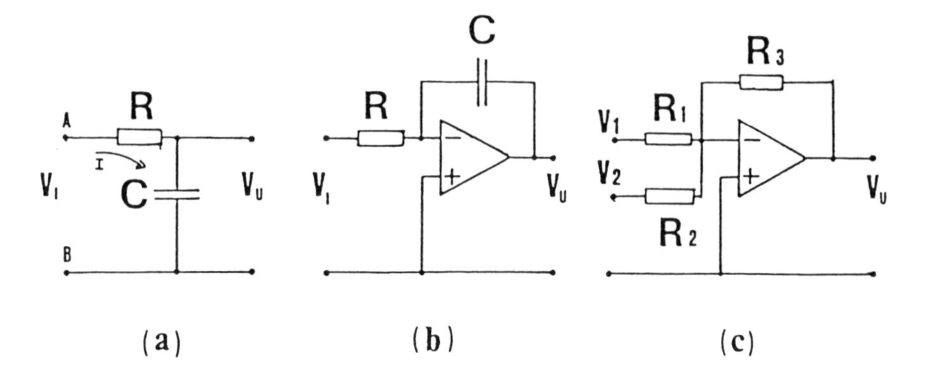

Figure 1

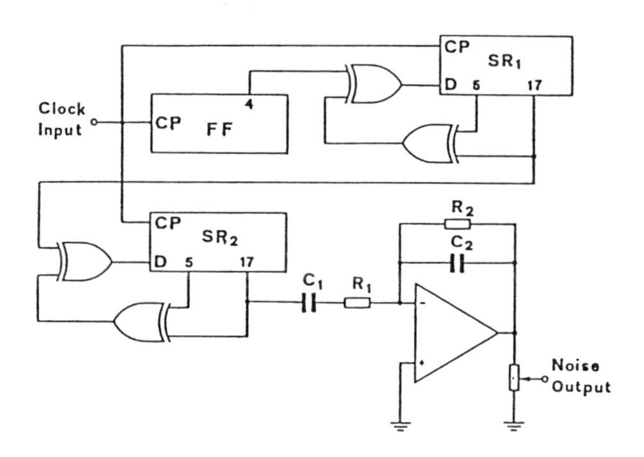

Figure 2

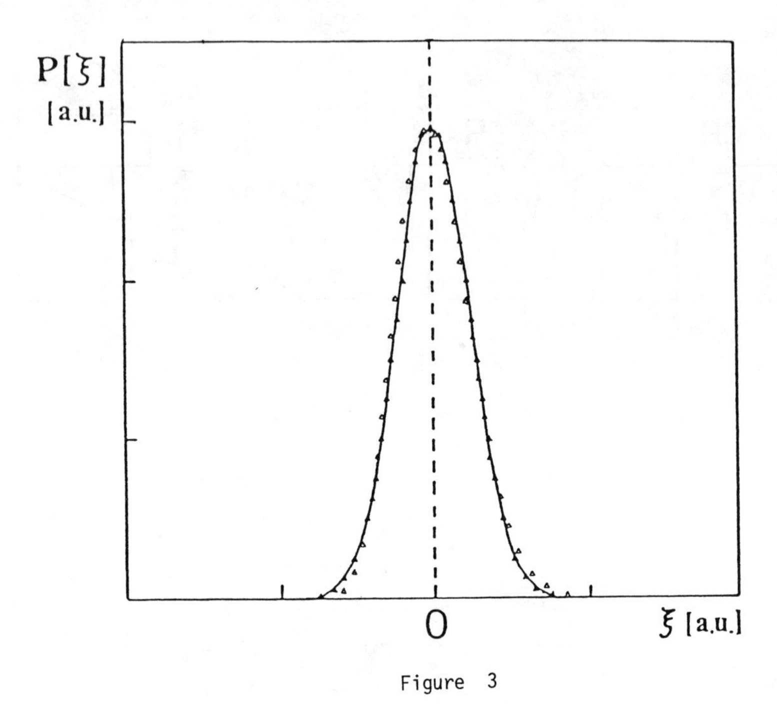

Figure 3

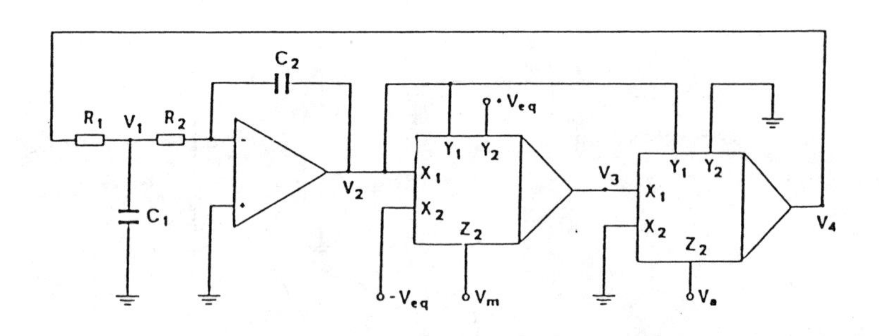

Figure 4

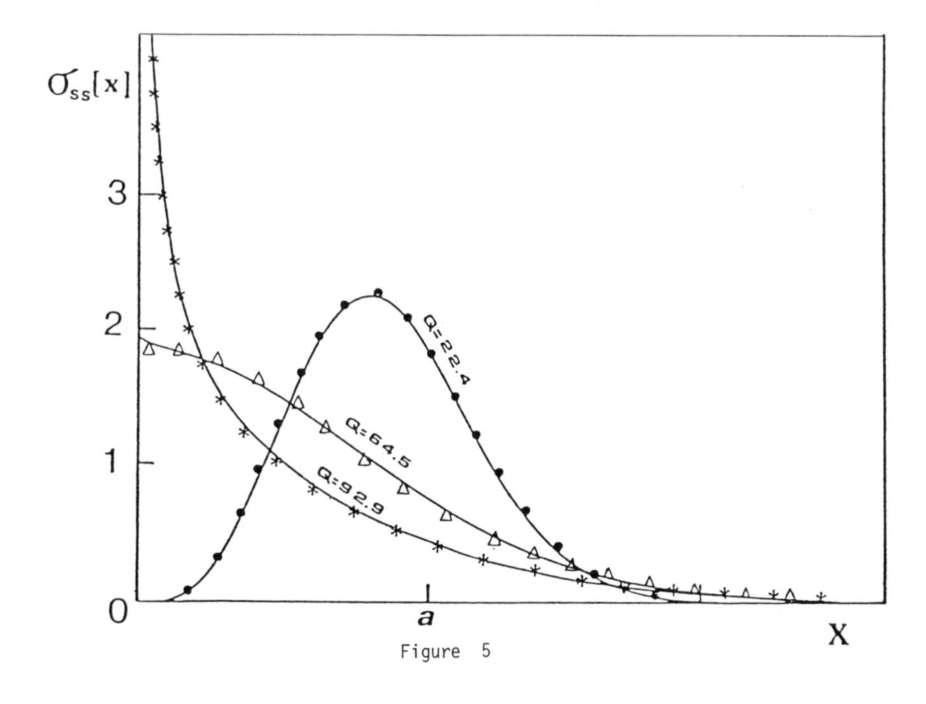

Figure 5

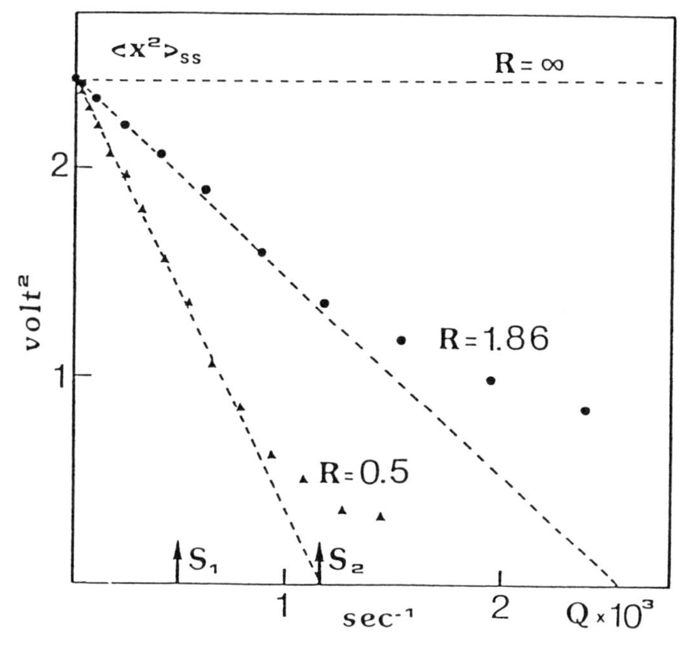

Figure 6

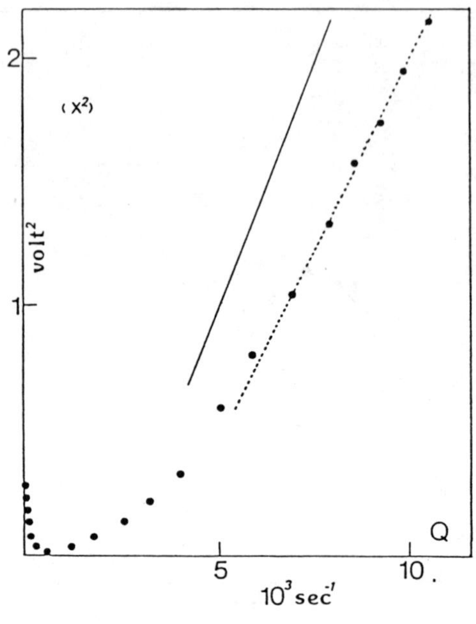

Figure 7

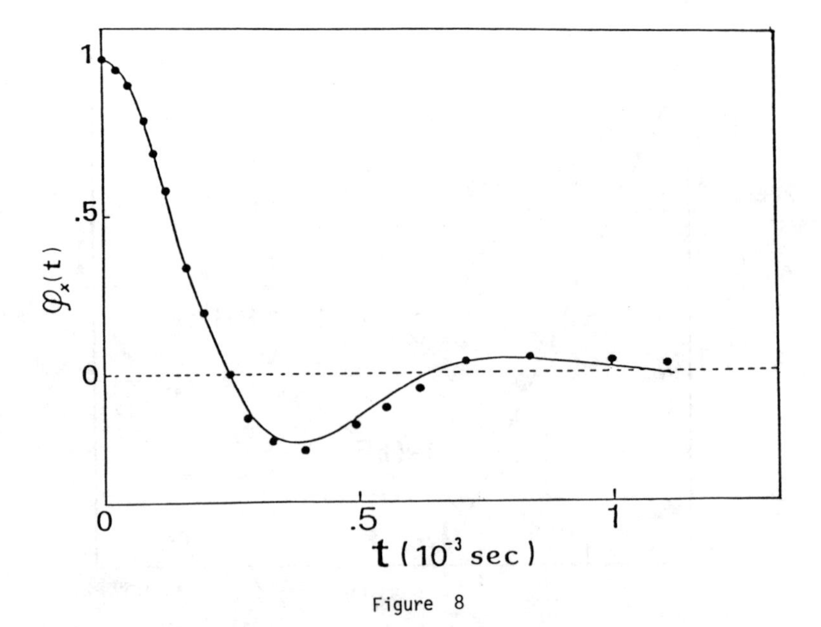

Figure 8

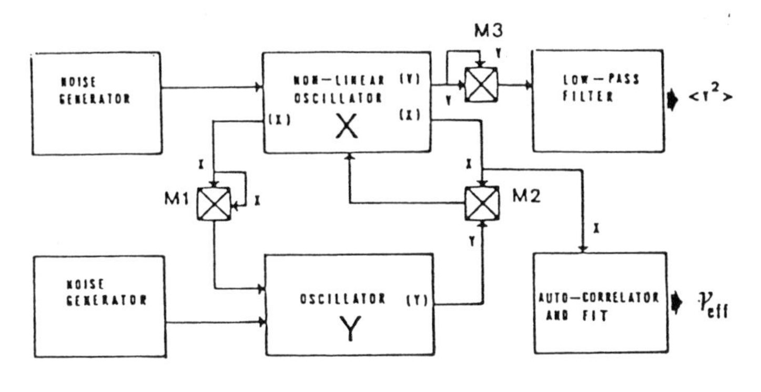

Figure 9

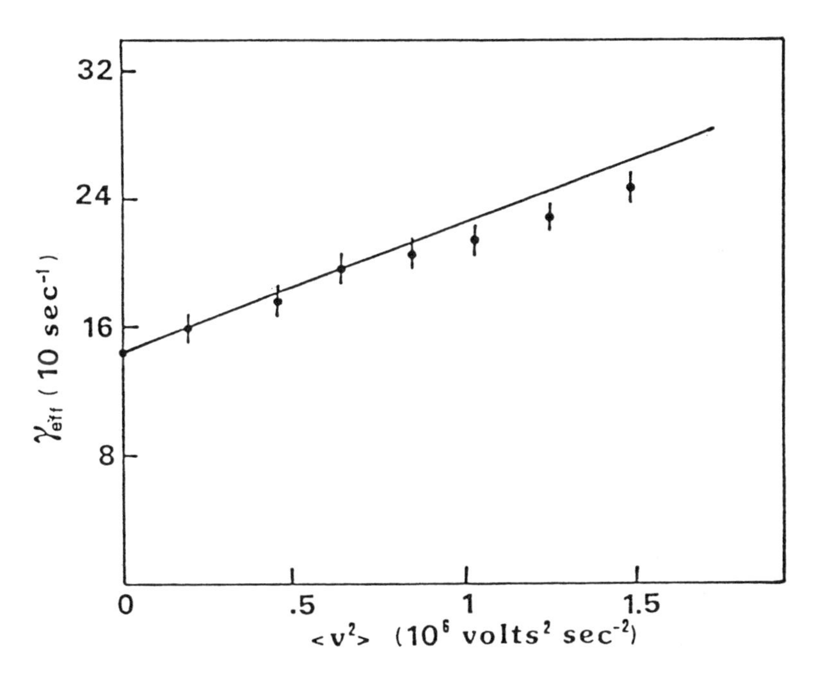

Figure 10

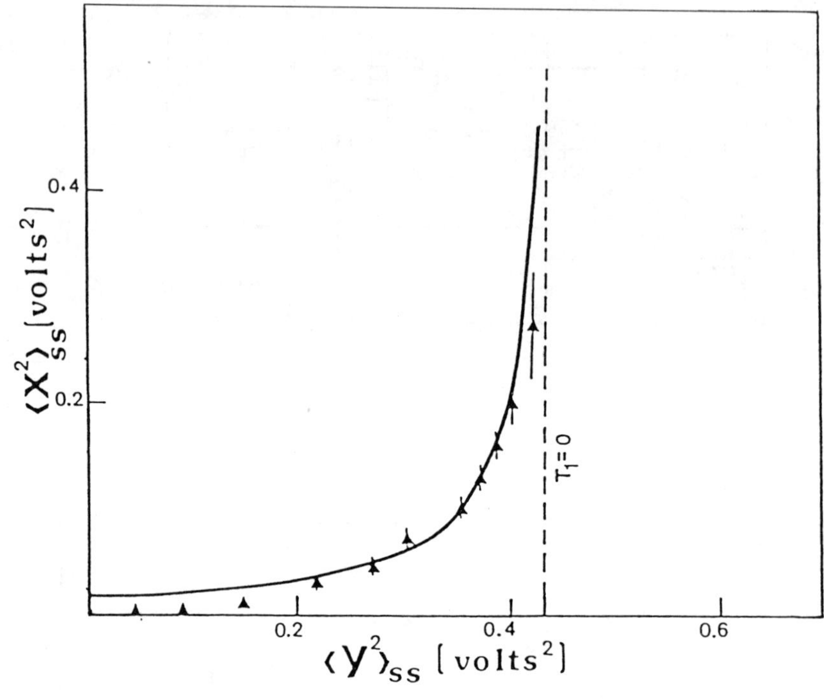

Figure 11

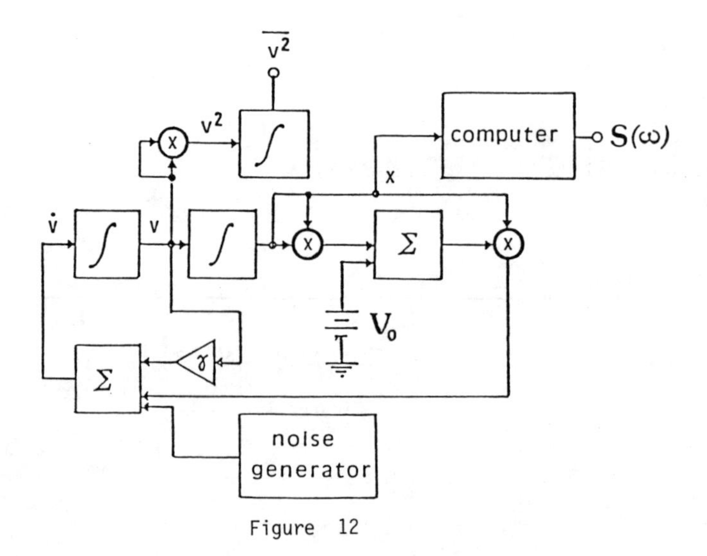

Figure 12

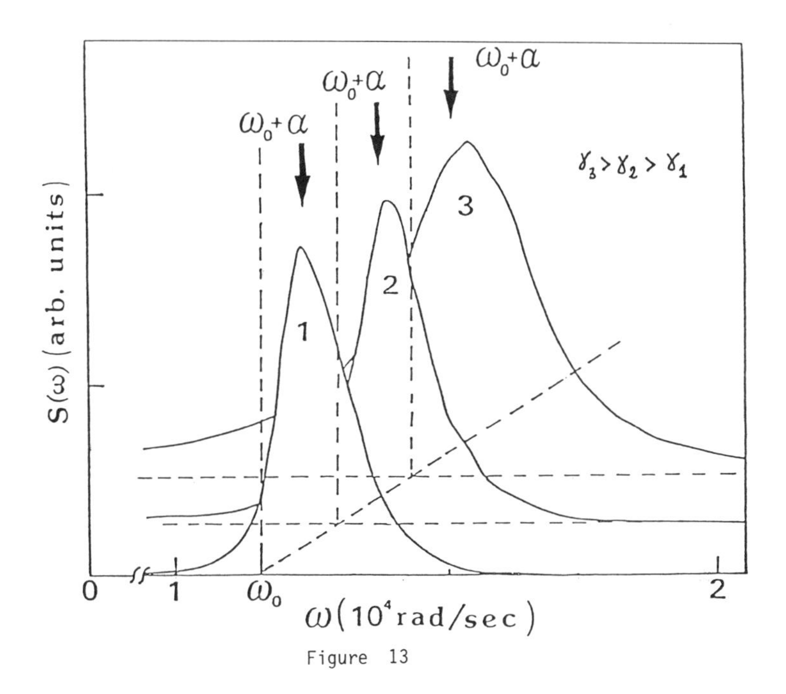

Figure 13

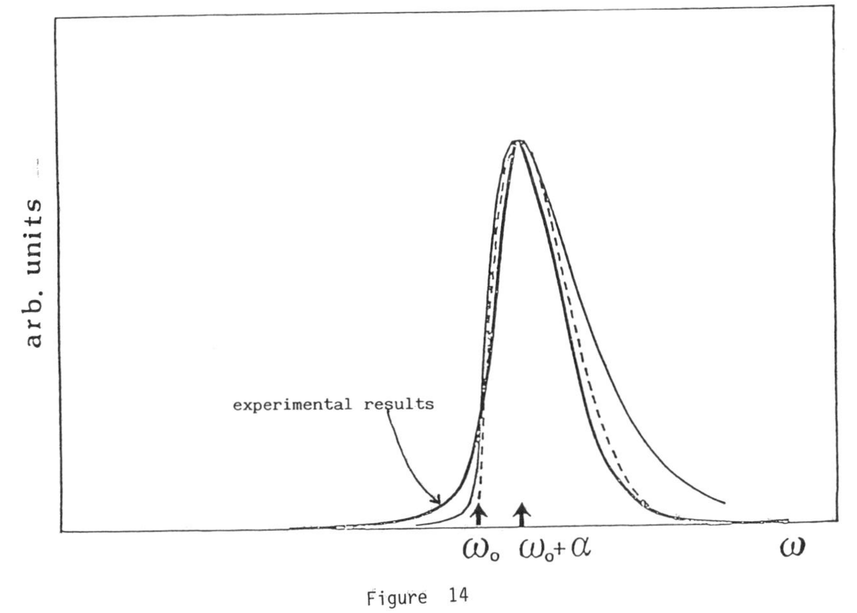

Figure 14

LIST OF PARTICIPANTS

- C. AGNES
  Politecnico
  C.so Duca degli Abbruzzi, 24
  10129 Torino

- T.F. ARECCHI
  Istituto Nazionale d'Ottica
  Largo E. Fermi, 6
  50125 Firenze

- A. BARACCA
  Dipartimento di Fisica
  Largo E. Fermi, 2
  50125 Firenze

- G. BARBAGLI
  Dipartimento di Fisica
  Largo E. Fermi, 2
  50125 Firenze

- G. BENETTIN
  Dipartimento di Fisica
  Via Marzolo, 8
  35131 Padova

- J.G. CAPUTO
  L.E.M.D. (C.N.R.S.)
  B.P. 166X
  25 Avenue des Martyrs
  38042 Grenoble (France)

- M. CASARTELLI
  Dipartimento di Fisica
  Via D'Azeglio, 85
  43100 Parma

- S. CILIBERTO
  Istituto Nazionale d'Ottica
  Largo E. Fermi, 6
  50125 Firenze

- D. ESCANDE
  Laboratoire PMI Ecole Polytechnique
  91128 Palaiseaux Cedex
  (France)

- S. FAETTI
  Dipartimento di Fisica
  Piazza Torricelli, 2
  56100 Pisa

- C. FERRARIO
  Istituto di Fisica
  Via Paradiso, 12
  44100 Ferrara

- F. FERRI
  Dipartimento di Fisica
  Via Celoria, 16
  20133 Milano

- V. FRANCESCHINI
  Istituto di Matematica
  Via Campi, 213 A
  41100 Modena

- L. FRONZONI
  Dipartimento di Fisica
  Piazza Torricelli,2
  56100 Pisa

- W. GADOMSKI
  Uniwersytet Warszawski
  Al. Zwirkiiwigury 101
  Warszawa
  (Polonia)

- L. GALGANI
  Dipartimento di Fisica
  Via Celoria, 16
  20133 Milano

- R. GIACHETTI
  Dipartimento di Fisica
  Largo E. Fermi, 2
  50125 Firenze

- P. GRIGNOLINI
  CNR
  Dipartimento di Fisica
  Piazza Torricelli, 2
  56100 Pisa

- G. LENZI
  studente della Facoltà
  di Scienze M.F.N.
  Firenze

- G.L. LIPPI
  Istituto Nazionale d'Ottica
  Largo E. Fermi, 6
  50125 Firenze

- R. LIVI
  Dipartimento di Fisica
  Largo E. Fermi, 2
  50125 Firenze

- G. LO VECCHIO
  Istituto di Fisica
  Via Paradiso, 12
  44100 Ferrara

- M. MIARI
  Dipartimento di Fisica
  Via Celoria, 16
  20133 Milano

- G. MORANDI
  Dipartimento di Fisica
  Via Campi, 213 A
  41100 Modena

- G.L. OPPO
  Istituto Nazionale d'Ottica
  Largo E. Fermi, 6
  50125 Firenze

- P. PAOLI
  Istituto Nazionale d'Ottica
  Largo E. Fermi, 6
  50125 Firenze

- M. PETTINI
  Osservatorio Astrofisico di Arcetri
  Largo E. Fermi, 5
  50125 Firenze

- A. POGGI
  Istituto Nazionale d'Ottica
  Largo E. Fermi, 6
  50125 Firenze

- A. POLITI
  Istituto Nazionale d'Ottica
  Largo E. Fermi, 6
  50125 Firenze

- A. POSILICANO
  Dipartimento di Fisica
  Via Celoria, 16
  20133 Milano

- A. PROVENZALE
  Istituto di Fisica
  Corso M. D'Azeglio, 46
  10125 Torino

- G. PUCCIONI
  Istituto Nazionale d'Ottica
  Largo E. Fermi, 6
  50125 Firenze

- N. RIDI
  Istituto Nazionale d'Ottica
  Largo E. Fermi, 6
  50125 Firenze

- G. RIELA
  Dipartimento di Fisica
  Via Archirafi, 36
  90123 Palermo

- S. RUFFO
  Dipartimento di Fisica
  Largo E. Fermi, 2
  50125 Firenze

- F. SIMONELLI
  Istituto Nazionale d'Ottica
  Largo E. Fermi, 6
  50125 Firenze

- V. TOGNETTI
  Dipartimento di Fisica
  Largo E. Fermi, 2
  50125 Firenze

- G. TURCHETTI
  Dipartimento di Fisica
  Via Irnerio, 46
  40126 Bologna

- C. VAIENTI
  Dipartimento di Fisica
  Via Irnerio, 46
  40126 Bologna

- A. VULPIANI
  Dipartimento di Fisica
  P.le Aldo Moro, 5
  00185 Roma

- B. ZAMBON
  Dipartimento di Fisica
  Piazza Torricelli, 2
  56100 Pisa